*Precognition
and Philosophy of Science*

An Essay on Backward Causation

Precognition
and the Philosophy of Science

An Essay on Backward Causation

Bob Brier

Humanities Press, Inc.
New York, New York

Library of Congress Cataloging in Publication Data

Brier, Bob.
  Precognition and the philosophy of science.

  Bibliography: p.
  1. Causation. 2. Cognition. 3. Science—
Philosophy. 4. Gale, Richard M   1932-
I. Title.
BD591.B77   133.8'6   73-12754
ISBN 0-391-00325-9

Printed in the United States of America

Contents

Acknowledgments

I should like to thank the members of the Philosophy Department at the University of North Carolina and my colleagues at C. W. Post for numerous helpful discussions of my work. Most of all I should like to thank George Schlesinger, my teacher, whose help was more than academic.

The publication of this book has been made possible by grants from the Parapsychology Foundation and the C. W. Post Center Research Committee.

Introduction

Parapsychology, as a new science, exhibits many problems which other sciences had in their infancies: the methods of gathering data are not very efficient; it is difficult to repeat an experiment and obtain the same results; and perhaps most important, there is no theory which explains the fragmentary findings of the investigators. Aside from these similarities between parapsychology and infant sciences, there is what may be a very important difference—parapsychological findings raise important philosophical questions. They suggest that the way we have been conceiving our (physical) universe may be wrong. Since there is a clash between the findings of parapsychologists and the conceptions and theories of philosophers, something has to give. Either new philosophical concepts must be formed to replace the old, or parapsychologists must admit that they could not have possibly obtained the evidence they believe they have—they made errors, fraud was perpetrated, the results are due to chance, etc. Since it does not seem credible that all the data accumulated by parapsychologists are due to error and fraud, philosophers should see what suggestions parapsychology offers as to what concepts might be altered and in what way.

In this monograph I hope to show how one concept, the notion of cause, might be altered in order to explain the phenomenon of precognition.

Since in any philosophical enterprise it is essential that there be no confusion of terms, it will be best if we first try to define some of the general areas of our topic and then select the specific area on which we will concentrate.

Several definitions of "parapsychology" have been suggested. They differ only slightly and I think it is reasonable to say that parapsychology is that branch of human inquiry which studies events which can not be explained in terms of physical principles. Some brief comments on this definition are necessary.

The tentative nature of this definition should be stressed. When it is said that the phenomena studied are not explainable in terms of physical principles, we mean that they can not be explained by the physics of today. Should new discoveries in physics yield explanations for what was previously thought to be unexplainable, the definition would have to be changed.

The phenomena studied by parapsychologists are generally divided into two major areas: extrasensory perception and psychokinesis. Extrasensory perception (ESP) is knowledge of the environment without the use of the senses. It occurs when a person knows something that is happening, will happen, or has happened, but this knowledge is in no way derived from his five senses.

ESP is divided into three categories: clairvoyance, telepathy, and precognition. Clairvoyance is extrasensory perception of physical objects or events that are presently in existence. In contrast to clairvoyance, telepathy is defined as the extrasensory perception of mental states or thoughts of other persons. Precognition is perception of an object or event which is in the future.

From the above it can be seen that ESP in its three forms involves the organism receiving information of his environment without the use of his senses. This is only one of the two areas which parapsychology studies. The other area involves the organism's influence upon his environment rather than the environment's influence upon the organism. This second aspect of psi is called psychokinesis (from the Greek ψυχη κινησις), or PK. This is the ability which has popularly been called *mind-over-matter,* and this is a fairly good description. PK is the direct influence exerted on an object by a person without any known intermediate physical energy or instrumentation.

The definitions offered above are admittedly sketchy and incomplete. It is hoped, however, that they will be sufficient for our purposes.

Of the three kinds of ESP, precognition is the one which raises most of the philosophical questions. If we want to try to explain how clairvoyance works we can imagine that some sort of as-yet-undetected wave emanates from the object clairvoyantly perceived and this wave is picked up by the person having the clairvoyant experience. With telepathy we can talk about brain waves from one person being picked up by another. We must grant that brain waves, as we now know them, are so weak that even with the most sensitive devices they can be picked up only a few yards from the body. Nevertheless, such stronger waves are *possible* explanations— they may some day be detected. How might we explain precognition? This is a different case. We find it difficult if not impossible to imagine waves that go into the future and return to the present bearing information about where (and when) they have been. How can events in the future, events which do not yet exist, make their presence known to those living in the present? This is the question around which this monograph will center. Can an event which is in the future have an effect which is in the present? This has sometimes been called the question of backward causation because the order of cause and effect is backwards. Normally the cause comes first and is followed in time by its effect. Here we are talking about a cause succeeding its effect.

Many philosophers have dealt with the possibility of an effect coming before its cause and almost all have argued that such a course of events is impossible. If this is true then this kind of explanation is not open to precognition and we must seek a different kind of explanation. In this essay I shall try to show that it is indeed possible for a cause to come after its effect, and consequently this is a possible way of describing what happens in a precognitive experience.

I

Gale on Backward Causation

Stranger: *Let us now reflect and try to gather from what has been said the nature of the phenomenon which we affirmed to be the cause of all these wonders. It is this.*

Socrates: *What?*

Stranger: *The reversal which takes place from time to time of the motion of the universe.*

Plato (*Statesman,* 270)

Gale has written extensively on the question of backward causation, and his approach is somewhat different from any other philosopher's. Consequently, it may prove instructive to begin by considering Gale's work as a unit in itself and trace the development of his arguments to his final position.

His first contribution in this area, an article in the *Review of Metaphysics,*[1] presents a basic attack which he maintains throughout his work on backward causation, although he gradually modifies it somewhat. Gale believes that it is an analytic truth that a cause cannot be later than its effect. However he correctly points out that if the evidence becomes terribly impressive, one may wish to give up an analytic truth.

---

1. Richard M. Gale, "Why a Cause Cannot be Later than Its Effect," *Review of Metaphysics* 19(1965): 209-34.

Gale first sets up the conditions for any example of backward causation which will make the reader relinquish the "analytic truth" that a person cannot influence the past if he is in the present. The strategy is to show that any example which satisfies all five of his conditions will require one to use words such as "intention," "memory," etc., in radically new ways, indicating the absurdity of talking of a cause occurring later than its effect. When Gale's later work is considered, these five conditions will have to be discussed, but since this is not necessary here, it will perhaps be best to postpone any discussion of them until later.

The example Gale offers involves what he calls retrowarning, that is, a case in which someone in the present warns someone in the past. Smith learns that at an earlier time Jones was in trouble and warns him by saying, "Jones, this is Smith warning you from the future." It is later learned that Jones had a paranormal experience and heard the words of Smith. Previously, Gale argued that one cannot know that what one is going to try to bring about has already happened, for this changes the meaning of "intend," "deliberate," etc. That is, if one is going to try to bring about a certain state of affairs, X, then it follows that he does not know if X will or will not be the case. This is so because he cannot deliberate about causing certain states of affairs which he knows already obtain or will obtain.

What Gale suggests is that, since Jones is in the past (and also his being saved or not being saved), then it is logically possible for Smith to remember that Jones was saved or not saved. Thus he could know that the event to be caused happened (or did not) independent of his actions causing the event in question. The contradiction derived is that it seems as if it is both logically possible and impossible for the agent to know the event in question occurred (or did not) independent of the intended action to bring it about.

It is evident that the basic form of Gale's argument is that of the *reductio*. When backward causation is entertained a contradiction of the standard form $(P \& \sim P)$ is derived. In this case P = it is logically possible for an agent to know the event in question occurred independent of the intended action to cause the event to have happened. The affirmative part of the contradiction offers no trouble. Where Gale goes wrong is in the derivation of the negative half of the contradiction.

Gale would be correct if he said it is logically impossible to deliberate about E if one knows that E has taken place. This, however, is quite different from saying that if it is logically possible that one know that E has happened, then it is logically impossible for him to deliberate bringing about E. Smith can intend to kill Jones even though Jones is already dead (but Smith does not know Jones is dead). It is surely not logically impossible for Smith to learn of Jones' death. Perhaps formalizing the argument will help.

q = knows of the event in question
r = deliberates bringing about the event in question

There are three propositions which are relevant here:

1. $[(\Diamond q) \& (r)]$
2. $\sim\Diamond(q \& r)$
3. $\sim[(\Diamond q) \& (r)]$

Proposition 1 is the positive half of Gale's alleged contradiction; it is granted a truth value of T. Proposition 2 is another true proposition concerning the situation presented by Gale. Proposition 3 is what Gale's contradiction requires (the negation of 1), but is one which is false. Gale has skipped from 2 to 3.

In private correspondence with this writer, Gale admitted his error and said,

> The source of my error is that I represent a conceptual truth of the form, 'p logically implies q,' as 'p implies necessarily q.' ...What I should have said is that it is necessarily true that if a person is deliberating about doing E then he does not (in fact) know that not-E.[2]

2. Letter to the author, May 10, 1968.

The force of this revision is worked out in a section of Gale's book, *The Language of Time.*

Gale begins the chapter entitled "The Impossibility of Bringing About the Past"[3] by discussing what he takes to be the various asymmetries between the past and future. After these preliminaries he points out that the central question will be whether there could be a description of a constant conjuction between events (such as is given in precognition experiments) which is of such a sort as to justify giving up the analytic truth that a cause cannot come after its effect. Gale's strategy again will be to show that, in giving up this analytic truth, certain neighboring concepts will also have to be radically changed. If it can be shown that many basic concepts will have to be altered, the suggestion will be that the analytic truth is conceptually central and not to be relinquished.

In the earlier treatment of Gale's work, his five conditions which any counter-stimpulation-example must meet were temporarily set aside. Since he also uses these criteria in his book, they will be considered more closely at this point.

1. The ordinary use of "past" and "future" must not be radically altered.
2. It must be the case that every occurrence of an event of type L in a situation of type S has been preceded by an event of type E.
3. L must be an intentional action.
4. E must not be a proximate or remote cause of L.
5. There must be no proximate cause of E that is earlier than L. [P. 112]

The reason for the first condition is almost evident. Without it, a counter-stipulation-example might fail through equivocation. For an example of an effect being past while the cause is present, etc., to be of philosophical importance,

---

3. Richard M. Gale, "The Impossibility of Bringing About the Past," in *The Language of Time* (New York: Humanities Press, 1968), pp. 103-33.

the words "past" and "present" must be used in their ordinary senses.

The second condition assures that L is, indeed, a cause.

The third condition states that it is necessary that L brings about E, or makes E happen. Gale tells us that the paradigm cases of causality are those in which an agent *intentionally* intervenes into some situation and brings about a change in it. Gale says that a paradigmatic case of causality is required because otherwise there would be no compelling reason for saying that L in S is sufficient cause of E, rather than E in S is a necessary cause of L. There are two things wrong with this line of reasoning.

The first mistake is that Gale is looking for compelling evidence rather than a logical possibility. That is, to show that backward causation is logically possible, it is not necessary to contrive a case in which it is the only explanation. All that is needed is to show that it is a possible explanation—one which might be accepted as the best explanation. It may be that, in every case where backward causation is possible, an alternative explanation is involved; consequently, there has to be a weighing of evidence to choose between the alternatives. However, backward causation is still a possibility.

A second question might be raised about this third criterion. Gale seems to think that, because L is ostensibly intentional, it could not have been caused by a prior event. At first this does not seem self-evident. Were this the case, the philosophical arguments concerning freedom of the will would be pointless. This is so because, if deliberative actions are not caused by anything prior, then there appears to be a freedom of the will, and the question is settled. He has offered no argument for establishing that E could not have caused L, an intentional act. This certainly seems to be a logical possibility, but Gale seems to feel he has compelling reasons for ruling it out.

The last two conditions involve the notions of remote

cause and proximate cause, both of which were explicated earlier by Gale.

> A proximate cause is a cause (in some undefined sense) which bears one of the following three temporal relations to its effect: (1) it can be simultaneous with its effect (to be called a 'proximate cause$_1$'); it can be temporally contiguous with its effect (to be called a 'proximate cause$_2$'); or (3) it can be temporally separated from its effect by initiating some process that in the normal course of things terminates in its effect (to be called 'proximate cause$_3$'). [P. 107]

> A *remote* cause of an event is a cause which, like a proximate cause$_3$, is temporally separated from its effect, but which unlike a proximate cause$_3$, is the proximate cause of some sequence of casually related events one of which is, in the context under consideration, said to be the proximate cause (in one of the three senses given above) of the effect. [P. 109]

Gale gives examples to help clarify what he means, but it is hoped that the above definitions will suffice for present purposes.

Condition 4 is that E must not be a proximate or remote cause of L. This is necessary because it is an analytic truth that causality is asymmetric; that is, nothing can cause its cause.

Condition 5 asserts that there must be no proximate cause of E which is earlier than L. If there were a cause of E prior to it, there would be no causal determining of the later event left over for L.

Gale uses this fifth condition to show how a counter-stipulation-example must fail. He first presents Dummett's example of the Indian braves on their six-day trip to prove their bravery.[4] The first two days they travel; the second two days they fight; and the third two days they return home. On all six days the chief dances to cause them to be brave. What is of special interest here is the chief's dancing on the last two days. Here the event to be brought about has already occurred or not occurred. The point Gale makes is that Dum-

---

4. M.A.E. Dummett, "Bringing About the Past," *Philosophical Review* 23(1964): 338-59.

mett says nothing about certain facts relevant to the warriors' being brave. That is, he does not discuss their training in fighting wild animals. This is certainly a possible cause of the warriors' bravery. Consequently Gale says, "On any occasion when the men's brave conduct can be satisfactorily accounted for by prior causes there would be no need to invoke the chief's dancing on the fifth and sixth days as the cause, or even a part of a cause" (p. 120).

Gale is correct in most of what he says about Dummett's example, but he is wrong in rejecting the example as a counter-stipulation-example. If one is considering the logical possibility of backward causation, all one should require from a counter-stipulation-example is a case in which backward causation is logically possible. Gale, by raising prior causes as *possible* alternatives is not arguing against the logical possibility of backward causation. To do that he would have to show the *necessity* of a prior cause, thus violating his fifth condition; all he has shown is the *possibility* of a prior cause.

Another way of putting this would be to say that Gale is looking for a logically possible case in which one *must* accept backward causation as the explanation. Bringing in the question of alternative explanations and the weighing of evidence is premature. It may well be the case that, whenever backward causation is logically possible, alternative explanations are available. But the discussion of which alternative is more plausible is quite a different question from whether backward causation is logically possible. This question will be taken up later when Mackie's work is discussed.

Gale next attempts to make a point which would be essential in ruling out all possible counter-stipulations. He is, of course, aware that one can merely stipulate in one's example that there is no prior cause of E. Gale claims, however, that all such examples in which Condition 5 is satisfied will cause one to depart from the ordinary concepts of a *causal explanation* and *law*. Gale points out that there are really two different interpretations of Condition 5: a weak one, W_5, and a stronger one, S_5. W_5 asserts that, every time an event of type L occurs, there cannot be a proximate cause of E earlier than L. S_5 stipulates that, every time an event of type E occurs, there is no proximate cause of it earlier than the subsequent occurrence of L. The relevant difference between the two forms of Condition 5 is that the weaker version allows for a case in which E occurs and there is a prior cause, provided that L does not consequently occur. Thus in W_5 E can exist without L.

One thing surprising is that Gale allows W_5 as a possible form of his condition. He is supposedly constructing an example in which L is the cause of E, but he seriously considers a case in which L does not occur. Surely this should be ruled out because it fails to fit the description of his example. In any event, Gale wishes to show that both W_5 and S_5, when satisfied, cause ordinary concepts related to *causation* to be destroyed. He begins with W_5.

Gale had previously argued that if C is sufficient for E, this will cover the period from the beginning of the cause to the end of the effect. This must be the case because in this period the presence or absence of some condition could prevent the effect from happening. It will first be noted that the meaning of the assertion is not very clear. That is, what does it mean to "cover a period?" One possibility is that if one asks, When was C sufficient for E? the correct answer is, From when C began to when E ended. But Gale does not state this. The next point to be made is that Gale's support for his assertion, whatever it means, is rather weak. He derives his assertion from the fact that some condition within the period in question could prevent the effect. Gale then uses the assertion to argue against W_5. However, without necessary support, he says, "The description of the situation in which one event is a sufficient proximate cause of another must not refer to what does or does not obtain at any time *earlier* than the *beginning* of the cause or *later* than the *end* of the effect" (p. 122). Apparently Gale feels this proposition concerning linguistic legislation is equivalent to his assertion about causes "covering" periods of time. This is not necessarily so. Surely it is possible for a cause to "cover" a certain period of time; and yet a description of the relevant circumstances can make reference to times prior to and following the period in question. If this is not the case, Gale certainly has not made it clear why it is not. In any event, Gale uses the formulation of the linguistic legislation to attack W_5. First he presents an appropriate translation for backward causation by replacing "earlier" with "later" and "beginning" with "end." Thus of backward causation, he says that the description of the situation in which one event is a sufficient proximate cause of another must not refer to what does or does not obtain at any time later than the end of the cause or earlier than the beginning of the effect. Then since W_5 refers to what does not obtain *earlier* than the beginning of the effect, it is rejected; and since W_5 is a necessary condition to be met by a counter-stipulation-example, all such examples must be rejected. Since Gale has not given any demonstration as to why his linguistic legislation is a necessary condition to be met by all examples of proximate causes, the force of his argument against backward causation is negligible.

If one substitutes S_5 for W_5, it would seem as if the difficulty mentioned above would be surmounted. This is so because S_5 does not refer to a time earlier than E. Since it asserts that there is no proximate cause of E other than L, it is, as a matter of fact, known that there is no proximate cause of E earlier than E. Gale says that S_5 does not really solve the problem because there is the question of whether there was an earlier proximate cause. The rules for use of "cause" in a counter-stipulation-example satisfying S_5 must be discovered.

Moving from his discussion of W_5 to S_5, Gale shifts ground, and this helps to indicate where he is confused. At first it would seem as if S_5 does assist the difficulty discussed. As noted before, Gale asserted that, "The description of the situation in which one event is a sufficient proximate cause of another must not refer to what does or does not obtain at any time *earlier* than the *beginning* of the cause or *later* than the *end* of the effect." This is then suitably altered for backward causation. Surely S_5, which is part of such a description discussed, satisfies Gale's condition. There is no reference to the time periods ruled out. However, Gale brings in other considerations. That is, he discusses possible uses of "cause" in such contexts and seems to feel this is the same as not satisfying his condition asserted. This strengthens the suggestion made earlier that Gale's undefended assertion is not a clear locution, and consequently whether or not the condition is satisfied (let alone justified) is not clear.

Gale has another line of arguing against backward causation. It involves the case of retro-warning and merits more serious consideration. Gale feels that he has revised his objections sufficiently to avoid the fallacy discussed before. In this case, all five conditions are satisfied: (1) The use of "past" and "future" is not radically altered; (2) L never occurs without being preceded by E; (3) L is an intentional action; (4) E is not a cause of L; and (5) E has no prior proximate cause and, in fact, no proximate cause prior to L. Given that the five conditions for a counter-stipulation-example are met, Gale now must argue that various conceptual changes are required. The argument is long, complicated, and invalid.

An important concept in Gale's thesis is *intentional* knowledge. (It will be remembered that L is an intentional act.) In the example, one can have intentions with respect to the past as well as the future. He can deliberate about causing Y to have happened as well as causing Y to happen. Gale asserts that one can also have *intentional knowledge* of the past as well as of the future and offers two definitions of intentional knowledge.

(i) An agent has intentional knowledge of one of his actions X if, and only if, he has an intention to do X and has non-intentional knowledge that he can do X.

(ii) An agent has intentional knowledge of an event Y which is not one of his actions if, and only if, he has intentional knowledge of one of his own actions X and knows that X is sufficient to bring about Y. [P. 125]

In the example, Jones can have intentional knowledge of L, his warning Smith, and also of E, Smith's being saved by the paranormal experience.

If an agent finds out either that he cannot carry out his intention to do X or that X is not sufficient for bringing about Y, he gives up his intentional knowledge claim.

It will be clear that intentional knowledge has a specific peculiarity. Intentional knowledge that Y will occur does not entail that Y will occur. This is so since at time T, one may have intentioned knowledge of Y, but later at T_1 he may have reasons for knowing not-Y. This will be important later.

Given the above, Gale will try to show that, in accommodating the notions of *backward intention* and *deliberation*, the concept of the *past* will have to be radically altered. He begins by presenting an analytic truth about the concept of the past.

An event is past if and only if it is logically possible for any person who is neither in doubt about the occurrence of this event nor believes that it did not occur to have a trace of it. [P. 126]

Gale tries to make clear what he means by "trace," but there are difficulties. He says that having a trace involves being in possession of a trace, and this trace enables one to know the event occurred. (Thus having a trace and being in possession of a trace are different.) Gale says that, on the basis of a trace, a person can make a non-inferential knowledge claim about what happened. This is puzzling. Normally, if one is shown a photograph of another's grandfather, on the basis of the photo he can *infer* that the grandfather had a mustache, was short, etc. Perhaps the reason that Gale says one makes non-inferential claims is that he is using "trace" as a success word. He says it is logically impossible for a person who has a trace of an event to either be in doubt or believe that it did not occur. Having a trace means more than having evidence; it entails knowing that the event in question occurred. This is somewhat strange. One can understand that there being a trace of an event entails that the event occurred. Here, however, having a trace entails knowing that the event occurred.

Gale's next step is to assert that "the only logically relevant reason for challenging or giving up a knowledge claim based on an ostensible trace is evidence which indicates that the ostensible trace is unveridical" (p. 127). Now Gale turns to his counter-stipulation-example.

Assume that Jones has seen Smith blown up, remembers it, and has other traces. Gale says that Jones is in a position to say that he knows E did not occur. This is certainly true, since according to Gale's discussion of trace, Jones *knows E* did not occur. Gale then says something which is surprising. "However, all of a sudden Jones begins to deliberate about whether to do L so as to bring about E in the past. At the moment Jones begins to deliberate he is logically required to give up his knowledge claim based on his ostensible memory. . ." (p. 127). This is so because it is not logically possible to deliberate on bringing about E and know not-E. Gale continues:

> What has happened is that Jones has had to give up his knowledge claim based on ostensible traces of not-E, not because he has discovered any new evidence to indicate that E did occur and that therefore these ostensible traces are unveridical, but merely because he has begun to deliberate about doing L so as to bring it about that E did occur. [P. 128]

This violates Gale's principle concerning the giving up of knowledge claims based on a trace. The agent has given up his knowledge claim for a reason other than one of the three Gale accepts.

Gale's description of the situation is not unusual. He

contrives a case in which an agent knows not-*E*, and then without reason begins to deliberate about doing *L* to bring about *E*. It is certainly true that at the same time the agent cannot know not-*E* and deliberate about bringing about *E*. What is unusual about Gale's description is his conception of the temporal sequence of things. He seems to think that first the fellow knows not-*E*, then begins to deliberate bringing about *E*, and this forces him to relinquish his knowledge claim of not-*E*. However, what must be the case is that first there is ostensible knowledge of not-*E*, then that "knowledge" is doubted, and then deliberation takes place. This is so because a necessary condition for deliberating to bring about *E* is nonknowledge of not-*E*. If an agent is deliberating to bring about *E*, this entails he does not know not-*E*. Gale seems to think it entails that the deliberating is *the reason for* giving up the knowledge claim. Not only is this an unjustified conclusion, it is logically impossible. A necessary condition for the deliberation is the lack of knowledge. The deliberation cannot be the cause of the lack of knowledge.

The case Gale has contrived is an impossible one. Jones could not merely begin to deliberate about doing *L* at time *t*, when at that time he knows not-*E* was the case. Thus it is not surprising that he is able to use it to show how the notion of *past* is radically affected.

Gale is aware of the objection just raised and feels that he has a reply to it. If one is to have a proper analogue to deliberation for backward causation, then one must allow for an agent to give up a nonintentional knowledge claim because he begins to deliberate about causing something to have happened. This is true, but in Gale's example, added to this stipulation is the restriction that the agent has knowledge of not-*E*. To consider the relevant future counterpart, one would have to stipulate that the agent has knowledge that not-*E* will occur. (It is not essential to know how he obtains this knowledge; he might have a paranormal experience; he might know that a sufficient condition for not-*E* obtains, etc.). Under such conditions it does not make sense for the agent to begin to deliberate about committing an act which will cause *E*. This is so since he *knows* not-*E* will occur. Thus the relationship of backward deliberation and forward deliberation is symmetrical with respect to being able to deliberate causing *E* when one knows not-*E* obtained or will obtain. Consequently if there is a radical change in the notions of *past, deliberation,* etc., it will occur not only for backward causation but forward causation as well.

Gale feels he has an answer to this reply also. First he asserts that, in an analogous case, the analogue to memory would be precognition, and since precognition is *logically impossible,* there is no specific proper analogue. Thus it is not merely contingent that there is no analogue to memory, and this is an argument against backward causation.

The first point to make is that in this example Gale does not rely upon memory but merely upon traces in general, of which memory is one. There might be a photograph of not-*E*, etc. Consequently it is not essential to bring in precognition in order to have a backward causation analogue. There might be traces that a future event will occur which do not involve precognition. Here one can consider the case in which it is known that a sufficient cause of not-*E* has occurred.

The second point to make against Gale is that his extremely brief argument against the logical possibility of precognition is not valid. Gale views precognition as an analogue to memory and says that, while it is possible for an agent to have memories without interruption of decisions he made as a result of past deliberations, it is not possible to have uninterrupted precognition of decisions one will make as a result of future deliberations. Were this the case, an agent might deliberate about an outcome he already knows—something which is not possible. Gale's point that one cannot have complete precognition and still have deliberation is well taken. However, neither can one have complete memory and deliberation with regard to causing an event to have happened. What Gale's thesis points out is that *complete* precognition and deliberation are incompatible. This does not show that precognition is not possible. Indeed, it does not even show that there is any one event which cannot be precognized. What it does show is that if a person precognizes an event and truly believes it will happen in the future, then he cannot logically deliberate about doing something to either bring about or not bring about that particular event. Indeed, it is even possible that one person can precognize that event and another can deliberate about causing or not causing it. This is similar to memory—one person can remember that an event occurred, and if the other is ignorant of this event, he can deliberate about causing it to have happened. Thus there does seem to be an analogue to memory-precognition—and these concepts relate similarly to deliberation.

Gale offers a very brief explanation as to why he feels *complete* precognition is required to have an analogue to memory. In the explanation he introduces the idea of "aforeknowlia," an analogue to amnesia. His one-sentence explanation is: "The reason for requiring uninterrupted precognition is that otherwise an agent could now precognize the future occurrence of *M* but subsequently suffer from aforeknowlia and then honestly begin to deliberate about whether to do *M*" (p. 131). What is not clear is what would be wrong with precognizing, then forgetting the precognition, and deliberating about doing *M*. There is no apparent logical contradiction involved, and it seems like an understandable and possible situation. A man might have a legitimate precognitive experience that he will be killed in a plane crash the next day. However, he is hit on the head and does not know he will be killed. He then discusses with his wife whether he should take the plane, even though from the fact that he precognized his death in the plane crash, it follows that he will, in fact, be killed in the crash. Gale seems to think there is something contradictory involved in such a situation.

One can construct a similar case with memory. A man

might remember that Jones was killed yesterday. However, now he is struck on the head and no longer remembers this. He can now deliberate about doing something to cause Jones to have been saved. The point where Gale may feel a logical contradiction is involved in such cases concerns the man's knowing the event in question occurred. That is, if he has a trace, this entails that he knows and consequently he cannot deliberate about causing or not causing the event in question. The point to be made here is that his memory (or precognition in the analogue) is the trace. Once he no longer has the trace, the entailment of his knowing of the event does not hold. (He may however still know the event occurred from other sources.) Thus there is no clash between knowing and deliberating.

Gale has still another argument which he feels is telling against backward causation. In this argument he attempts to show how the notion of *experimentation* is destroyed when one admits backward causation as a possibility. When one wishes to experiment to see if X under circumstances S causes Y, then one performs a simple experiment. First one ascertains that S, then does or brings about X. The crucial part of the experiment is the observation to see if Y occurs. This is how one determines if X causes Y.

In the case of backward causation, there is a somewhat different situation. Here if one wants to experiment to see if X causes Y to have happened, he can bring about X when S obtains, but he cannot now directly observe Y, which is in the past. All he can obtain now is indirect evidence of Y. Thus with the backward causation case, one cannot verify the causal claim in the same way he can verify the causal connection claim in the case of forward causation. Indeed, since Y is in the past one can better verify whether it occurs before the causal action in question is performed! All there can be is a mock experiment.

From the above it seems as if Gale has one basic point which he supports in two ways. The basic point is that one cannot (in the usual sense of "experimental") experimentally verify the claim that a cause succeeded its effect. This is so because one cannot directly observe the effect after the cause has occurred, and because no real experiment is possible, since one can better verify if the effect occurred *before* the cause has occurred.

Concerning Gale's first approach, this seems clearly false. Direct observation is not a necessary condition for an experiment to determine if there is a causal connection between two kinds of events. In science one tests hypotheses by gathering evidence, and there is no restriction on the kind of evidence which is relevant, which rules out indirect observation. When one sees a trace on a cloud chamber this is evidence that an electron was there. Surely Gale has no valid reason for ignoring such evidence.

Gale's second approach is related to the first, and consequently the objection to it is somewhat related to the reply given above. In a causal experiment, what is being tested is not whether the effect occurs, but whether, if the cause occurs, the effect occurs. Since the hypothesis involves a relation between two events, one a cause and the other the effect, one cannot test the hypothesis until the time periods for both the cause and the effect have elapsed. That is, in the case of backward causation, the hypothesis cannot be tested until the cause has been brought about. Thus waiting until the cause occurs is not merely a mock experiment. It is *the* experiment. In the experiment Gale describes, given that the effect in question occurs, it is crucial to ascertain that the cause in question also occurs.

A different way of arguing against Gale's second point is to show that the situation is analogous to forward causation experiments. He is correct in pointing out that in a backward causation experiment, at the time of the cause, one no longer has the possibility of directly observing the effect. However, in forward causation experiments, at the time of the effect, one no longer has the possibility of directly observing the cause. It is true that in the forward causation experiment one can first directly observe the cause and then the effect, but in a backward causation experiment one can directly observe the effect and then the cause. In both experiments both events in question can be directly observed in temporal sequence. The only difference is that the order in which they can be observed is reversed; but this is precisely the difference between backward and forward causation.

Gale has one last argument against backward causation. Again he selects a concept and attempts to show how it is radically altered by admitting backward causation as a possibility. This time the concept is *punishment.* In this case one is to assume that E is an event which is morally reprehensible but has no other cause than Jones' subsequent action of doing L. Here since the illegal event is the result of Jones' intentional action, one can hold him responsible and punish him accordingly. But he is being punished for something he has not yet done. The contradiction involved here is that the notion of punishment involves reforming the criminal so that he will not perform such acts in the future. If the punishment is a success with Jones and he does not perform L, then he has been punished for a crime he did not commit.

There are numerous ways of arguing against Gale on this point. It is hoped that one will be sufficient. First one might point out that Gale's example, as he presents it, involves a contradiction which is not the result of the introduction of backward causation into the picture. Gale begins by stipulating that L is the cause of E. He then permits Jones to be reformed so that he does not do L. How, then, is it the cause of E? Surely the problem here is that Gale has allowed Jones to be punished before he has committed the crime. This would not be the case with backward causation. Given that E has occurred, one would hope that there would be a search for the culprit who caused E. Until there is adequate evidence someone caused E, then one would not expect to punish anyone. Here it might be rightly said that given E, one can infer *L*. This is true, but it could be that mere inference that a punishable act will occur is not sufficient reason to punish someone. It might be argued that a necessary condition for justly punishing someone is that he has committed the act for which he is being punished. Thus it is not true that, if backward causation were the case, people would have to be punished before they committed their crimes. Here too evidence would be required, and this might involve waiting until the cause occurred.

If Gale's argument were valid, it could be suggested that there is an analogous situation for forward causation. One might imagine the case where psychology is sufficiently advanced so that, given the initial boundary conditions of an individual, under limited circumstances, his future actions can be predicted. Let it also be stipulated that there is a specific case in which it is known from predictions that the fellow (if he is not restrained) will commit a crime. Here one is faced with a "problem" quite similar to that raised by Gale with respect to backward causation. Thus even if his argument were valid, it would not be telling against backward causation.

II

Dummett and His Critics

Now, of course, I mean by effect that which has already come into existence and has been completed by the activity of these faculties—for example, blood, flesh, or nerve.
Galen (*Natural Faculties,* Bk. I, Ch. 2)

Dummett's example of the dancing tribal chief was discussed in the treatment of Gale. The example is a central one in the backward causation question, and indeed Dummett is a philosopher of central importance in the controversy. Because Dummett's contributions in this area form a coherent picture of the development of several lines of attack on the problem, if his work is discussed chronologically, the various arguments that have been presented will first be seen, then the respective replies.

Dummett's first article on backward causation resulted from a symposium (held at a meeting of the Aristotelian Society) entitled "Can an Effect Precede its Cause?"[1] In this article he discusses the Humean view of causation—that is, "cause" is synonymous with "sufficient condition." In this view, however, backward causation seems to be a possibility, since the relation between necessary and sufficient conditions

---

1. M.A.E. Dummett, "Can an Effect Precede its Cause?", *Aristotelian Society Proceedings* supp. 27(1954): 27-44.

can hold between an earlier and later event as well as between a later and earlier event.

Generally, a distinction is made between remote and proximate causes. A remote cause is one which comes temporally before its effect and instigates some process or event which later causes the effect in question. All remote causes are causes of proximate causes which are simultaneous with their effects. Problems with each of these kinds of causes have arisen. With remote causes, it seems absurd to say that causes precede their effects. If the effect does not take place immediately, what does finally make it take place? Further, if there is temporal distance between a cause and its effect, then it seems as if there is never any certainty that the effect will take place. This is so because, after the cause, one can intervene and assure that the effect will not occur.

Dummett presents solutions to this dilemma by making observations on the way causation is viewed: A cause operates upon something, and once it stops operating, the thing goes on in the same way until some other cause operates upon it. However, what "goes on as before" may be a process which terminates in the effect. Thus there is no temporal lacuna, as one can always retrace a causal chain from an effect to its remote cause.

The problem with proximate causes is not so easily disposed of and is one which is most relevant to the backward causation problem. Since proximate, or immediate, causes are always simultaneous with their effects, how does one decide which of the two events is the cause and which the effect? Since there is an answer to this question (one does, in fact, make such decisions), this might at first suggest that temporal priority is not an essential feature in the notions of causality. When one is confronted with cases of simultaneous causes and effects, he calls the one which can already be accounted for (without reference to the other) the "cause" and the other, the "effect." Thus he does decide the question without reference to which is earlier and which is later.

It might be argued that deciding which of the two events can be causally accounted for already makes use of temporal direction. That is, one looks *backward* in time for a cause of either of the two events in question, and this is the decision procedure. This is true, but it is not damaging to the backward-causation theorist, for he can here raise the relevant question, Why is it not possible to look forward in time for a cause of either of the two events? Thus there is still no reason for rejecting backward causation as a logical possibility.

Using the notions of *remote* and *immediate cause*, Dummett presents a very forceful argument against backward causation. Any cause is either remote or immediate. Since they are simultaneous with their effects, immediate causes present no case for backward causation. Only remote causes remain. For a cause to be remote, it must be the immediate cause of the beginning of some process which continues and eventually is the immediate cause of the effect in question. Dummett then says this shows that a remote cause cannot come after its effect. "A remote cause can be connected to its remote effect only by means of a process which it sets in motion, i.e., which begins at the moment it operates and goes on after that. . ." (p. 31). In a sense, by definition Dummett has ruled out backward causation. This he does in the last two words of the sentence quoted ("after that"). That is, what would be required in a case of backward causation would be a remote cause which instigates a process which has gone on before that instigation—a process going backward in time rather than forward. Since this is what is at issue, Dummett should not rule it out as he does.

Since Dummett has rejected the possibility of a cause coming after its effect, it is surprising that the remainder of his article offers much support to the backward-causation thesis. This is so, however, because he next turns his attention to what he calls "quasi-causal" explanations. (It will be seen that there is little discernable difference between a cause and a quasi-cause.) In these cases it might be observed that an event is a sufficient condition for the occurrence of some earlier event. To have such a quasi-causal explanation these conditions would have to be met:

(1) The earlier event, which is to be explained by the later event, would have to be incapable (as far as can be judged) of being explained by a simultaneous or earlier event.

(2) The earlier event must not be the remote cause of the later event.

(3) It must be possible to give a causal account of this later event which does not contain a reference to the occurrence of the earlier event.

If these three conditions were satisfied and there was evidence for the constant conjunction of the two events, this backward quasi-causal explanation would have to be accepted. Dummett offers an example which seems to satisfy the above three considerations.

A man always wakes up three minutes before the alarm of his clock goes off. Frequently he does not know whether the alarm is set, and if so, for what time. An instance might even be stipulated in which the man wakes up and, for some irrelevant reason, a friend walks into his room and sets off the alarm exactly three minutes later. In such a case it might be said that the man wakes up because the alarm clock is going to go off.

Dummett points out two limitations involved in most quasi-causal explanations. First, unless one can specify the process initiated by the cause which eventually results in the effect, one will not have much of an explanation. Second, it is generally preferred that causal laws connect as wide a range of phenomena as possible. It would be desirable to say that events of the same kind as A cause events of the same kind as B. In the example, he says, neither of these two conditions is satisfied.

It might be the case that in backward causation there is difficulty in tracing the process. With the case of the alarm clock, one might try to envision an echo from the clock going backward in time. There would be difficulties however. One might trace the sound from the time the alarm goes off at t_{10} to t_9, t_8 ... t_0. But this would involve hearing the alarm at these times as well as when the man wakes up. Thus the situation could be described as: the man hearing the alarm; the man waking up and continuing to hear the alarm up until t_{10} + the time it takes to shut off the alarm. The question might also be asked, When does the sound stop, and if it ceases to be audible when the man wakes up, why at that time? These are merely examples of the kinds of problems which arise when trying to specify the process in backward causation. Here, Dummett seems to have a point when he says there will not be much of an explanation. However, there are also cases of forward causation in which one cannot specify the process initiated by the remote cause but is still willing to call the relationship causal. Thus a causal connection is not ruled out merely because there is difficulty in connecting two events.

The matter concerning the possible generality of a backward causation law is not easy to settle. It is difficult to specify criteria for generality, but just why Dummett asserts why such laws cannot cover kinds of causes and kinds of effects is not clear. It is conceivable that under certain conditions, S, whenever A occurs B occurs, and B is prior to A. Thus events occurring under condition S might be said to form a kind, and the backward-causal law would appear to be general. From the discussion of the two restrictions Dummett places on his quasi-causal explanation, it would seem as if they do not have to be the impoverished explanations Dummett feels they must be.

There is, however, a compelling *a priori* argument against the possibility of backward causation. Dummett expresses it as follows:

> ... to suppose that the occurrence of an event could ever be explained by reference to a subsequent event involves that it might also be reasonable to bring about an event in order that a *past* event should have occurred, an event previous to the action. To attempt to

do this would plainly be nonsensical, and hence the idea of explaining an event by reference to a later event is nonsensical in its turn. [Pp. 34-35]

What is now left for Dummett to do is to explicate precisely why it is absurd to do something in order that something else should have happened. To do this he presents his example of the magician, and uses this example to test several arguments.

A magician has a spell for causing the next day's weather to be clear. Whenever he recites the spell, he subsequently finds out that on the next day at the place mentioned in the spell the weather is good. There arises an occasion on which the magician wishes the weather to have been good yesterday at a certain place. Not knowing what will happen, he recites his spell, putting yesterday's date where he usually recites tomorrow's. Later he finds out that there was, in fact, good weather. Further, every time he tries his spell in this manner, he learns that there was clear weather.

One way of illustrating the *a priori* error involved is to entertain the backward causation possibility, utilizing the distinction between the absurdity of doing a certain kind of action and the absurdity of describing a situation in a certain way. In backward causation what is absurd is the way the situation would have to be described. With forward causation it can be said that A is brought about in order that B should occur. In backward causation it must be stated that, by seeing if B could be brought about, whether A occurred is being discovered.

Dummett is quick to point out that the magician would not say that the reciting of his spell is *finding out if the weather was fine*. One uses such a description only when there is no question of bringing about, and all that is left is discovery. This is not the case with the magician. Thus this argument is surely not sufficient to rule out backward causation *a priori*.

The second *a priori* argument against backward causation

(or a quasi-causal explanation as Dummett prefers) involves the magician's not being able to recite the spell. Either the weather was fine or it was not. If it was fine (and if the weather having been fine is a necessary condition for his reciting the spell), then it follows that he cannot help but say the spell. But to say this is to deny that the spell is a sufficient condition of the weather's having been fine. That this argument is not compelling can be seen from the fact that an analogous attack against forward causation can be made: Either an event will or will not take place. If it will take place, then one cannot help but perform the action in question, but this is tantamount to admitting that the action cannot be a sufficient condition for the future event.

Dummett has an interesting suggestion as to why the argument is fallacious. A brief statement of that suggestion is all that is necessary here. There is an inappropriate use of a contra-factual conditional. When one is concerned with a regularity which works counter to ordinary causal regularities, the usual methods of deciding the truth of a contra-factual conditional break down. Such considerations will be dealt with in a later section.

After rejecting the two *a priori* arguments against backward causation, Dummett concludes a general discussion of the problem with an example intended to suggest that there is no absurdity in accepting the possibility of regularities such as the ones discussed.

Assume a man discovers that every time he says "Click!" before opening an envelope, he finds that it never contains a bill. He does this for several months and unaccountably never receives a bill. Concerning this state of affairs Dummett concludes, "Nothing can alter the fact that if one were really to have strong grounds for believing in such a regularity as this, and no alternative (causal) explanation for it, then it could not but be rational to believe in it and to make use of it" (p. 44).

Flew Contra Dummett

The other symposiast on the backward causation problem is Flew, and his contribution is published by the Aristotelian Society in the Proceedings immediately following Dummett's paper.[2] Flew represents a rather different approach, and his paper may rightly be called an attack.

Flew begins by pointing out that it is conceivable that there might be possible species of causes—that is, by changing one of the features normally associated with causation, there might still be an analogous notion. What is at issue here is whether the temporal properties of cause and effect can be exchanged so that an analogue rather than something totally different is left.

An obvious approach—one which Dummett rejects—is to say that, if the cause came after the effect, then it would be the effect and not the cause. This begs the question and Flew argues that it is too facile. There is a more important difference between the two concepts, and this difference entails the temporal priority of the effect over the cause. The crucial difference is that causes *bring about* their effects. Given this, any attempt to discuss effects prior to their causes must "either be frustrated by contradictions; or end in a fraud. [The contradiction is] that what has been done might be undone" (p. 48). The fraud would involve calling something a cause which could scarcely be called an analogue, since it could not involve *bringing about.* Flew is correct with respect to one horn of the dilemma: it would be a fraud to call something a cause which does not bring about an effect. Flew seems mistaken, however, with regard to the other horn: with backward causation one is not committed to saying that the past can be undone. This would be nonsense. Rather, here one can readily make use of the distinction

---

2. A.G.N. Flew, "Can an Effect Precede its Cause?" *Proceedings of the Aristotelian Society* supp. 27(1954): 45-62.

between changing the past and affecting the past. One cannot change the past or undo what has been done. Rather, what is at issue is whether one can affect the past; that is, by a present action cause something to have happened which would not have happened otherwise. This does not seem to entail a contradiction. Thus Flew's first attack on Dummett must be rejected.

In his brief analysis of Hume's pioneering work on the notion of cause, Dummett points out that, since sufficiency seems to be the Humean criterion for causality, there is nothing to rule out a cause coming after its effect. This is merely the noting of an obvious point. This observation, however, causes Flew to discuss at considerable length the deficiencies in the Humean view. Flew's main objection is that Hume's discussion of causality is from an *observer's* rather than a *participant's* point of view, and this leads to overlooking an important aspect of causality. Because an observer merely notes the constant conjunction of *A* and *B*, he is not confronted with the temporal aspects of causality. Were Hume dealing with the problem from a participant's point of view, he would have observed that *A* can be used to bring about *B*. This notion of *use* (*power, effectiveness,* etc.) in bringing about might have led Hume to the realization that causes must come before their effects. Thus rather than accept the Humean thesis, the fact that it permits backward causation should cause one to re-examine the thesis.

What Flew is saying is not a very serious criticism of Dummett's position. Dummett merely says that Hume's position (be it correct or incorrect) does not raise serious objections to the backward causation thesis. What Flew points out about the Humean point of view is interesting, and probably correct, but this is not relevant to Dummett's position, since Dummett was not arguing that Hume was correct (and consequently deriving the possibility of backward causation).

Flew next turns his attention toward Dummett's example of the alarm clock and suggests alternative explanations. The first is coincidence. This is certainly a logical possibility, no matter how strong the evidence. But Flew goes on to say that one can consider the correlation between the man's waking and the alarm going off as coincidence "yet not dismiss it, but use the slug-abed's waking as a sign of the imminent sounding of the alarm" (p. 56). When one grants that an observed connection between two events is a coincidence, one also grants that one does not expect the correlation to hold up in the future to the same degree. This is part of the meaning of "coincidence." If a person guesses cards randomly chosen from a deck of playing cards and is 100% accurate, it might be said that his success was due to coincidence. If this is said, however, it entails that his future guesses are not expected to be 100% correct. Further, it would not be expected that his guesses could be taken as signs of what the cards would be. Thus it is surprising to see Flew both advocating the coincidence hypothesis and suggesting that the man's waking be used as a sign that in three minutes the alarm will go off.

It might be asked, How, on Flew's account, does one distinguish between coincidence and constant conjunction (i.e., cause in general)? It is clear that such distinctions are made with forward causation and coincidence. Here, such distinctions are often based on the repetition of the conjunction of the events. If it happens once, it may be coincidence. If the later event always follows the earlier, the coincidence-connection hypothesis will be rejected for a stronger kind of connection. With backward causation, there is no need to postulate a different decision-making procedure.

Flew's second attack on the alarm clock example involves Dummett's second criterion which any case of quasi-causal explanation must meet. This criterion he presents in two ways which he believes are equivalent: ". . . there would have to be reason for thinking that the two events were not causally connected; i.e., there must be no discoverable way of representing the earlier event as a causal antecedent (a remote cause) of the later" (p. 32). The first thing that comes to mind is that the two formulations of the criterion are not equivalent. The first formulation seems difficult to satisfy, since it requires evidence ("reason for thinking") for a negatively stated hypothesis ("not causally connected"). The second formulation is perhaps preferable, if for no other reason than it avoids the problem involved in the first. Here again, however, there might be trouble, not in satisfying the criterion, but in knowing when the criterion has been satisfied. What is required is that there be no *discoverable* way of viewing the earlier event as the cause of the later. Here, even though such a relationship may not have been discovered, one might not know of a relationship which is discoverable, but as yet undiscovered. Perhaps a combination of the two formulations might better express Dummett's intention and be more reasonable in terms of being satisfied: there must be no reason (apart from temporal order) for thinking that the earlier event is the cause of the later one.

It might be argued that whenever the alarm clock and the awakening are conjoined, the situation can be viewed as the awakening causing the alarm to go off, as well as the alarm causing the awakening. To this objection it might be said that what is being argued is merely the *possibility* of backward causation, and this condition seems satisfied by the example. This, however, only postpones the philosophical showdown. It can then be said that what is meant by *logically possible* is that, in some conceivable situation, backward causation is the best explanation. Thus, although in every case of backward causation there may be alternative explanations, what is necessary for the backward causation thesis is to present an example in which backward causation is not only a possible explanation, but the best alternative.

What is at issue here is really the explication of what is involved in the notion of *logical possibility*. The above argument asserts that logical possibility involves the best alternative in some possible world. Against this thesis one might point out that, traditionally, logical possibility merely involves a lack of contradiction. Thus if no contradiction can be drawn by entertaining the idea of backward causation, then it is a logical possibility.

Further, even in the explication involving possible worlds, the position presented is unorthodox. Here, logical possibility has traditionally been interpreted as *true* in some possible world, while the above position interprets it as entailing the *best available evidence* in some possible world. That the two positions are not identical should be evident, since false propositions are often believed on the basis of available evidence. A complete discussion of this topic would involve what is entailed by the notion of *truth,* and this is clearly a topic beyond the limited scope of this monograph.

Flew's objection to Dummett's second condition is that it would be almost impossible to grant that it has been satisfied, since one would not know where to decide to give up the search for the forward causal connection. The reformulation of the criterion takes care of this objection, since all that is required is that at any given time a forward causal connection has not been found. Surely only such a tentative position can be required from science.

The third but not last objection to the alarm clock illustration is not a clear objection. Flew points out that the experimentalist testing the causal connection will be in a peculiar position. If he has a quasi-effect, he must try to stop the quasi-cause from occurring. If he succeeds, then it was not really a quasi-effect. Only if he observes the effects and continually fails to stop the causes will they seem genuine. Why Flew considers this description an objection is not clear. He makes no attempt at deriving a contradiction, and it does not seem as if there is one to derive.

The fourth objection is a verbal one and asserts that any explanation in terms of *causes* "in any new sense of the word" will not be precisely what is desired since it will only be "a substitute carefully wrapped in the old verbal pocket."

Flew seems to be ruling out any possible solution to the problem, since he seems to be rejecting "any new sense of the word." Surely if one wishes to change a feature of a concept (but still have an analogue) one will have to consider the result, or new sense of the word. What should be the issue is not whether there is a new sense of the word, but whether there is an analogue. Flew seems to have collapsed the distinction between a new sense of a word and an analogue.

Finally, Flew's last attempt to show Dummett's example to be absurd is not an argument, but an assertion of what is to be demonstrated. One sentence is offered: ". . . any attempt to suggest that pseudo-causes *operate,* that is, are (and could be used as) levers to control the past, as real causes operate, that is do (and can be used to) bring about effects, must be radically absurd" (p. 57).

To summarize, Flew offers five objections to Dummett's proposed case of backward causation, and none of them is a compelling reason for rejecting the example. If one is to reject the possibility of backward causation and pronounce Dummett to be in error, it will have to be on ground other than what Flew has presented so far.

Flew's concluding remarks are full of sound and fury and bear the appropriate significance. For this reason it will not be profitable to go into a detailed discussion of the points raised. It will perhaps be better to single out one correct (but trivial) charge Flew makes against Dummett. It will be remembered that Dummett argued that, if a man "knew" what would happen, anything he could do with respect to the occurrence of the event in question would be both fruitless and redundant. Flew correctly points out that one may know that he will do a thing, but this does not make his future actions either redundant or fruitless. The reason Flew's objection is trivial is that when discussing such matters, it is understood that the subject is nonintentional knowledge. All the examples Flew presents deal with intentional knowledge of future actions.

Black Contra Dummett

The article by Dummett is a pivotal one in the backward causation controversy perhaps because it raises and discusses so many of the issues involved. Consequently, it is not surprising to find that Flew is not the only one commenting upon Dummett's early work. In his article, "Why Cannot an Effect Precede its Cause?"[3] Black enters into the arena. The first half of this paper discusses the question in general, but the latter half deals specifically with Dummett's paper. Because this section is primarily concerned with Dummett, this writer will first begin with the comments on Dummett, then show how Black's general comments are merely being applied to Dummett's example.

Black begins with his discussion of Dummett's alarm clock example. His first point concerns the cause of the alarm going off. It is known that the cause must be other than the waking up of the man, for this would be a case of forward causation. If the causes of the alarm going off are discussed, this will involve the mechanisms of the clock, and there will be little difficulty in doing this. However, this also shows that regardless of whether the sleeper has awakened, the alarm can be either caused to go off or prevented from going off. One might wait for the man to wake up and then immediately destroy the clock. Here, the waking could not be the effect of the alarm going off. If it is impossible to either make the alarm go off or prevent it from going off, an explanation of the causal mechanism of the clock proves to be insufficient, and it might be necessary to say that the waking is the cause of the alarm going off three minutes later. Thus it seems as if there are contradictions involved in backward causation, and the conditions Dummett set down for a case of quasi-causal explanation are inconsistent.

---

3. Max Black, "Why Cannot an Effect Precede its Cause?" *Analysis* 16(1955-56): 49-58.

At this point one might turn to the first part of Black's paper to see his more general comments. By analyzing these comments, it will readily be seen how they led to his criticisms of Dummett, and the reader will be in a better position to evaluate Black contra Dummett. First, Black presents an example which is somewhat analogous to Dummett's.

The example is of a man who can always predict (A) the final outcome of the tossing (T) of a penny. Black claims to have demonstrated the logical impossibility of a cause preceded by its effect, so he must now offer an alternative way of describing the situation. He considers the case where A and T have a common cause (X) which is prior to both. This would mean that whenever X occurred, both A and T would follow. But this does not allow T to occur in the absence of A. Thus T must have a sufficient cause which is not a sufficient cause for A. Consequently, this answer is not possible.

Black's next consideration is that T and A are caused by X and Y respectively, that X and Y are prior to T and A and causally independent. This can describe the situation, but is not satisfactory. What one is thus committed to say is that the perfect correlation of T and A is merely coincidence. If Black wishes to base his position on the philosophy of language, he must consider what is meant by "coincidence." In the case of a coincidence, one does not trouble oneself with looking for a causal or lawful explanation. Thus if Smith and Jones are both in New York on the same day (and both have never been there before), and a person is told that it is merely a coincidence, then he does not look for an explanation for their joint presence in New York. This is included in the meaning of "coincidence." To say that the situation Black has outlined (perfect prediction of coin tossing) does not require an explanation is a position which one doubts Professor Black would wish to defend.

Black has reduced the situation to what he describes as follows:

The final description of the hypothetical case is, accordingly, as follows: When Houdini is asked under hypnosis the question "How will the penny fall?" there is no earlier cause for his saying heads rather than tails. When the penny is tossed a minute later and agrees with Houdini's answer, there is a sufficient cause for that outcome of the toss, the cause in question being causally independent of the previous answer. [P. 54]

Black says that it can still be shown that the effect has not preceded its cause—one can wait for A, the prediction, and then prevent T. Though not directly relevant, one suggestion is offered here. It is an empirical question whether or not T can always be prevented, one for which Black has no right to assume an affirmative answer. This may seem curious at first, but may be illustrated by one case of a seemingly spontaneous precognition in which the event precognized could not be avoided:

... I was serving in a western desert, and I vividly dreamed that I was in Salisbury station and my wife was not there to meet me. I should explain that I had only been married three months prior to leaving the Colony, and the subject of my return was a constant one in our letters. I wrote to my wife and told her of the dream, and she wrote back a letter in which we both, so to speak, had a good laugh. However, by a set of most extraordinary circumstances entailing the flaunting of a number of precautions taken against the actuality, the dream actually occurred in real life some time after that. These circumstances included the catching of dysentery, having the convoy broken by an enemy submarine, and so arriving unheralded, and the arriving in South Africa and leaving again upon a Sunday so that I personally was unable to send a telegram, and finally the loss of the telegram by the Sergeant to whom I had entrusted it who was staying on the port of arrival.[4]

By this example it is not suggested that T of Black's example cannot be prevented. Rather this case is offered merely for its suggestive value: It is an empirical question whether T can be prevented. Surely it would be agreed that in most cases it seems as if certain things can be prevented

---

4. Louise E. Rhine, "Precognition and Intervention," *Journal of Parapsychology* 19(1955): 15-16.

from happening—like the tossing of a coin (perhaps). However, Black is too quick to generalize. (To be fair to Black, one should point out that in cases where it is impossible to prevent T, it might be said that A is a sufficient cause of T, and thus it is not a case where the cause succeeds the effect.)

In these limited cases one must reject backward causation, but as was seen earlier, such cases are not exhaustive of those possible.

Black's second, and similar, point—that one cannot only abstain from tossing the coin but can toss it in a manner so that it will disagree with A (if heads is predicted, toss it so that tails comes up)—suffers from the same deficiencies as abstaining from tossing.

The more relevant and stronger criticism here is that, when Black considers both the cases where T is prevented and T is arranged to contradict A, he no longer is examining his example. This is surely not a case of precognition as Black claims it is.

Chisholm and Taylor have a different argument against Black's thesis.[5] Their tactic is to show that if Black's argument is valid, similar problems arise with regard to the future. That is, a situation can be imagined in which it is known (by some means) that a future event E is going to happen. One can then arrange for C (a cause-to-be of E which is also future but prior to E) not to happen. This, according to Black's reasoning, would show that C cannot be regarded as a cause of E. This is so because an event which can be prevented from occurring cannot be regarded as causing an event that in fact does occur. Thus if Black's argument shows that no cause can follow its effect, similar reasoning would seem to show that no cause can precede its effect.

From the above, it can be seen that Black has failed to

5. Roderick M. Chisholm and Richard Taylor, "Making Things to Have Happened," *Analysis* 20(1959-60): 73-78.

demonstrate the "logical absurdity" of claiming that an effect can precede its cause. It should also be obvious that the example he offers is quite similar to Dummett's alarm clock illustration. When Black discusses arranging the toss to contradict the prediction, it is precisely the same (and involves the same problems) as preventing or causing the alarm to go off.

Black concludes his article with some objections to Dummett's example of the envelopes having no bills when "Click!" is uttered. Since these arguments are the same as those offered against the alarm clock example, there is little point in discussing them.

Flew Contra Black and Dummett

Soon after Black's paper appeared, Flew published a rejoinder to it.[6] In this paper Flew states that Black's paper is basically a restatement of the ideas in Flew's earlier paper, with one exception. Flew admits that Black has added an argument of his own (the coin tossing example), but that Black's rejection of this example as a possible case of backward causation is in error. Here Flew makes an observation noted before in the discussion of Black's paper. When Black requires that the coin toss can be arranged so as to contradict the prediction, he is adding a stipulation which is unwarranted. He supposedly is examining cases of ostensible precognition, and surely a case involving incorrect predictions will not do.

Since Flew believes that Black's error would not have been made had he read Flew's earlier paper more carefully, the majority of this second paper is restatement of what has been discussed earlier. Consequently, attention can now focus on a second, and very important, paper by Dummett.

---

6. Antony Flew, "Effects Before their Causes? Addenda and Corregenda," *Analysis* 16(1955-56): 104-10.

Dummett and the Dancing Indian Chief

In the chapter dealing with Gale's work on backward causation, the now famous example of the dancing Indian chief was briefly discussed. This example was first presented in a 1964 article by Dummett.[7] In this paper Dummett presents refinements of the position he presented earlier and makes some interesting additions to his thesis.

He begins by stating that causality is associated with one direction rather than another, not merely because of semantics, but because there is an objective asymmetry in nature. Obviously thinking of Flew's earlier objection, Dummett goes on to say that this causal asymmetry would reveal itself if people were merely observers and not agents. A person can conceive of reversed causal processes which he views, having no power of his own to affect them. However, once he places himself in the picture as an agent who can act within the causal framework, it seems as if problems arise. "But the conception of doing something in order that something else should have happened appears to be intrinsically absurd: it apparently follows that backward causation must also be absurd in any realm in which we can operate as agents" (p. 340). It might first be suggested that Dummett has prepared a most unusual argument. That is, it would seem that, if it is absurd for an agent to try to bring something about, this absurdity arises *because* backward causation is an impossibility, and not the other way around. It could also be said that the assertion of this "intrinsically absurd" action is an assertion of precisely what is at issue and what is to be demonstrated. Of this Dummett is aware, for in the remainder of his paper he attempts to explicate how this absurdity arises.

Dummett first makes it clear that one cannot change the past, so any attempt to do this will indeed be absurd. He

---

7. M.A.E. Dummett, "Bringing about the Past," *Philosophical Review* 23(1964): 338-59.

points out that Jewish theologians believe it blasphemous to pray to God to change something in the past, since this is logically impossible. If this is what retrospective prayer comes to, there is a similar situation with regard to prayer for the future. God cannot cause something to happen which will not happen. With the case of future prayer, one is asking God to cause something to occur at a future time which would not have occurred without His action. Retrospective prayer can be interpreted in a similar light; one can say that he is asking God to cause something to have happened in the past which would not have happened had He not acted at some later time. This is one familiar case of *affecting* the past.

The case discussed involves God, omnipotence, etc., and thus may be a special case from which one would not want to draw any general conclusions. This is so because in this case God, Who knows everything—past and future— thus knew at the time of the crisis that the future prayer would be made; so He might grant the prayer. Here the retrospective praying makes sense. What one must look for is a case in which it makes sense for a human to try to affect the past without Divine help or involvement. One immediate solution is to say that man can have foreknowledge. Thus there can be situations analogous to the case of God, situations in which one does something now so that someone in the past could have had foreknowledge of this action and could have acted accordingly. There are, however, more general lines of attacking the problem, and it may be more fruitful to pursue these.

Dummett makes an important comparison between one argument against backward causation and the argument for fatalism. It might be said that either a man's son has been killed or has not; whatever the man does now is pointless. Similarly, the fatalist argues either the son will be killed or will not; thus any present action will be pointless. Generally the fatalistic argument is rejected, so an important question, and an interesting way of approaching the problem, is to ask, Is there any refutation of the fatalistic position for which a parallel refutation of the backward causation thesis cannot be constructed?

It is first necessary to consider the fatalistic position. The fatalist moves from "He will not be killed" to "If he does not take precautions he will not be killed." This seems legitimate. Next, from "If he does not take precautions, he will not be killed," the fatalist infers "His taking precautions will not be effective in preventing his death." This does not seem permissible. It seems possible for the precautions he takes to be instrumental in averting his being killed. This, as Dummett knows, is a very brief sketch of a refutation of fatalism, but it will perhaps be sufficient for the purpose here. It must now be seen if there is a parallel refutation for the absurdity of affecting the past. It is here that the case of the dancing Indian chief comes in.

As was seen before, the young warriors of a tribe travel for two days to a mountain; stay on the mountain for two days and fight animals to prove their bravery; then travel homeward for two days. Their chief dances on each of the six days in an effort to make the warriors demonstrate their bravery. Then an observer reports to the chief whether the men were brave. The question now is, How can the chief be convinced of the absurdity of his dancing on the last two days, after the young men have either demonstrated or not demonstrated their bravery?

The chief might be told that either the men have been brave or have not been brave; consequently, anything that he does on the last two days will be superfluous. Here the chief could reply that, merely from the fact that they have either been brave or not been brave, it does not follow that what he is going to do will not be effective in making them to have been brave. That is, it may be precisely because he is going to dance that they have been brave; otherwise they would not have been. From his reply, it can be seen that the chief is replying in much the same way as Dummett did when countering the fatalistic argument. (It can also be seen that the chief is a regular reader of *Analysis*.)

The chief might be shown that, after the observer returns and tells him the results, he would not think of dancing. To this he can say that, since he knows the result, there is no point in dancing. However, although the chief's knowledge may affect whether he thinks there is any point in performing the dances, it cannot make any difference to the effect the dances have on what has happened. Thus the chief might be challenged to dance when he already knows the men were cowardly. Here, either he dances and thus shows the dancing not to be a sufficient condition of the bravery, or he finds himself unable to dance. Thus the bravery might be the causal condition for his dancing. In the chief's defense, Dummett points out that it is not fair to say that either the chief's behavior is a sufficient condition for the men's bravery or it is totally unconnected. There may be a strong enough correlation between dancing and bravery and nondancing and cowardice to make it reasonable for the chief to continue the ritual, even though there are a few cases when he finds himself inexplicably unable to dance, etc.

By allowing the chief this out, it seems as if Dummett has weakened considerably this case being argued. That is, the emphasis has shifted from showing the nonabsurdity of believing a present event to be a sufficient (causal) condition of an earlier event to showing the chief to be nonirrational in continuing his dancing on the last two days. These are two quite different positions.

If the chief agrees to the experiment in which he is to try to dance after he has heard the observer's reports, then there is another possible outcome. Each time the chief attempts to dance, he succeeds and later learns that the observer was mistaken, had been bribed, or was lying, etc.

From the discussion of the possible outcomes of the experiment, it seems as if the chief can maintain two beliefs which have frequently been asserted to be incompatible: (1) there is a positive correlation between his dancing and prior bravery; and (2) the dancing is something within his power to do. However, Dummett points out that a third belief, usually taken for granted, is incompatible with the above two: "... that it is possible for me to find out what has happened (whether the young men have been brave or not) independently of my intentions" (p. 357). This third belief must be rejected because no matter what evidence the observer presents, the chief dances and new, stronger evidence appears. This situation is quite the same with respect to the future. One never believes (1) an action is positively correlated with a future event; (2) it is in his power to perform that action; and (3) he can know independently of his intention to perform the action whether the future event will occur. Dummett makes the point that a difference between the past and future is that for any past event one normally feels that it is, in principle, possible to have nonintentional knowledge of the event; but this is not the case with the future. Here it might be noted that Dummett has earlier ruled out foreknowledge, which would again make the picture a symmetrical one.

It might be argued that the entire notion of evidence is threatened by the possibility of backward causation. This is so because any evidence may be rendered misleading in a world in which backward causation can at any later time be introduced to make events—which according to existing evidence have not happened—to have happened. It is true that any evidence may later be discovered to be misleading, but this does not mean backward causation is impossible nor even that the concept of evidence is destroyed. Even without backward causation, any evidence is open to contradiction by evidence which later turns up. Thus there is, in principle, no different effect on the notion of *evidence* if backward causation is allowed than if it is ruled out. Cases can even be considered in which, when there is misleading evidence, backward causation occurs, and the truth is still undiscovered because the backward causation does not lead to the turning up of new evidence.

Since the three beliefs mentioned are incompatible, at least one must be abandoned, and as Dummett suggests, one *could* choose to abandon the third and retain the first two. It might be pointed out here that Dummett may have presented the third belief too strongly. It may not be that it is "impossible" for the chief to learn of the cowardice or bravery of the men independently of his intentions. It may just be that *in fact* the chief did not learn of what happened independently of his intentions and dancing. There are other considerations which can be raised concerning the inconsistency of the three propositions, and these will be saved for when Gorovitz's relevant comments are discussed.

From the above it will be seen that no argument has been presented which will make the chief relinquish his belief in backward causation. Rather he can choose to change his beliefs about the acquiring of knowledge of past events.

Gorovitz on Dummett's Later Position

The issue of the *Philosophical Review* which contained Dummett's article also contained a paper by Gorovitz, commenting on Dummett's thesis.[8] One point on which Gorovitz focuses his attention concerns whether temporal asymmetry, or even merely the notion of causality, would reveal itself to someone who was only an observer of the world and could not interact with it. Dummett suggested that such a distant observer could discover the temporal asymmetry in the world, and Gorovitz argues against this. Since this point is primarily a psychological one and not in the mainstream of what is being discussed, this issue can be bypassed for one

---

8. Samuel Gorovitz, "Leaving the Past Alone," *Philosophical Review* 23(1964): 360-71.

which is more closely related to the backward causation question.

In his article, Dummett claimed that it is possible to conceive of a world in which "everything happens in reverse." Here Gorovitz points out that Reichenbach[9] has presented a demonstration that temporal order is reducible to causal order. This comes from relativistic considerations. Another question which would arise is, If everything (including clocks) runs backward, how would one discover this? Here it should be pointed out that, even if Gorovitz and Reichenbach are correct in suggesting as a possibility a world in which "everything runs backward," there is still the important question of a world in which there are *some* instances of backward causation. It is to this possibility that attention should now be turned.

There are three propositions which are crucial to Dummett's analysis of causation. His assertion is that the following three propositions are inconsistent for an agent in a world with intentional actions:

(i) There is a positive correlation between his doing A at time T and the occurrence of event E at time T_1 prior to T.

(ii) Doing A at T is within his power if he so chooses.

(iii) It is possible for him to find out whether E occurred at T_1 independently of his intention to perform A or not.

Before going into the argument it will perhaps avoid confusion if the impact of iii is clarified. At first it looks very similar to the crucial assertion made by Gale, and one might wish to say that all three propositions are compatible, just as in Gale's example there was only the illusion of a contradiction. In this case, for the three propositions to be inconsistent, one would need for iii an assertion that the agent has

9. Hans Reichenbach, *The Direction of Time*, ed. Maria Reichenbach (Berkeley and Los Angeles: University of California Press, 1956), p. 25.

found out whether E occurred at T_1 independently of his intention to perform A or not.

Here there would be a clash with the notions of *deliberation, intention,* etc. However, from the manner in which Dummett argues, it is clear that he does not mean *actually find out,* rather that there is a *logical possibility* of finding out. Because Dummett believes that the conjunction of i and ii clashes with iii, there is a choice as to which is relinquished, and it is only prejudice which causes i or ii and not iii to be abandoned.

There is, however, an important difference between the way Dummett sets up the problem and the way Gale set it up. Gale was talking about a case in which the agent actually performs the action in question. Dummett's ii is not explicit and merely talks about the possibility of the agent acting. That is, what is asserted by ii is that it is possible for the agent to do A at T. Also, iii asserts only a possibility—that it is possible for the agent to learn what was the case. Thus only i of the three propositions deals with an actuality. It is a fact that A is positively correlated with E. Given that ii and iii involve modal operators, the question becomes, Just where is the inconsistency? It is not between ii and iii, since there is no problem in it being possible for an agent to know of the event in question and possible for him to bring it about. This was seen from the discussion of Gale. The inconsistency does not arise between iii and i, since all there is here is the positive correlation from the past and a possibility of finding out whether E. By conjoining i with ii, nothing is gained in the way of contradicting iii; thus the three propositions are consistent.

Because the three propositions are consistent, it is unfortunate that Gorovitz accepts Dummett's claim that they are not, and then goes into a discussion of the logic involved in Dummett's argument. There is no point in dealing with this misguided section of Gorovitz's paper; rather it will be more profitable to review some of his more interesting arguments against backward causation.

Gorovitz presents an interesting example involving the possibility of shooting oneself in one's past. For example, given that a marksman can fire a bullet in 1968 and kill someone in 1944, a problem arises. What if he were standing next to that man killed? It seems as if the bullet cannot be two feet to the left of him. If it were, it would kill the marksman, and this is impossible, since it is certain he is alive in 1968. How can such a restriction on the path of the bullet be explained? Normally physical objects like guns are thought of as being capable of shooting either of two men standing side by side. Similarly, such peculiar limitations are not ordinarily imposed on volition. One normally can choose to shoot either of two men. Surely such changes in the notion of physical objects and volition suggest that there is something wrong with talking about affecting the past.

There is a reply to this objection. First, it must be granted that for a man to fire a shot, it is a necessary condition that he was not killed prior to the time of the firing. Given this, the reply is evident. There is no curious anomoly; a bullet could have killed him in 1944, but surely not one fired by himself in 1968. The killing of him in 1944 by a bullet (any bullet) removes a necessary condition for him to do any subsequent shooting. The argument would run something like this:

(1) To fire a shot in 1968, a man must not have been killed prior to 1968,

(2) A man fired a bullet in 1968.

(3) The man was not killed prior to 1968.

Now what is left to remove is the seemingly curious finding that a bullet cannot stray two feet.

Actually, it should be said that the bullet *cannot have strayed,* and perhaps this removes some suspicions. Consider the case of a reporter who was on a plane which crashed. All but the reporter were killed. He is now writing the story of the incident. Here too, it is the case—*he could have been killed,* but he was not. However, *he cannot have been killed;* he is now writing the story. The situation is quite similar to the one discussed in which a person fires a bullet. That bullet *cannot have killed him;* he has just done the shooting.

Thus the conclusions are that it does involve a logical contradiction to say that it is possible to kill oneself in the past, but this is not the kind of contradiction which is damaging to the backward causation thesis.

Gorovitz's last argument against the possibility of backward causation involves a proposed clash with the notion of *memory*—he claims that Dummett has failed to discuss important possibilities; for example, the case in which the chief is a witness to the warriors' fighting the animals. Here the chief himself observes the bravery or nonbravery of the warriors. If the men are brave, the chief will observe this before his final two days of dancing. Thus says Gorovitz, he has no need to dance. (It might be pointed out that, as Gorovitz sets down the conditions, there is not even an ostensible case of backward causation. Consequently, it is not of interest to the discussion.) It is important to note that Gorovitz in this instance makes no use in his argument of the fact that the chief observes the warriors. That is, Gorovitz could just as easily say that there is no need for the chief to dance if the warriors were brave and the chief were not present. If he has any argument at all, then it would hold in all cases of proposed backward causation. That is, since the effect occurs before the cause, then Gorovitz might for any case say that the cause is not necessary. But this is just what is at issue: whether a necessary condition for an event can occur after that event has happened. Thus this approach is begging the question.

There is also the possibility that the chief's subsequent failure to dance will cause, after all, the braves to have been cowardly. Gorovitz points to the chief's memory as the problematic area. That is, how can one account for the error in memory in such a situation? Further, after the chief dances, will he retain the same memory?

Here a point made earlier is relevant. In the backward causation claim, it is not asserted that the past has been changed, but that it has merely been affected. Gorovitz says, "Or will his memory abruptly change at such time as the warriors change from having been brave to having been cowardly?" (p. 368). He seems to feel his example calls for the past being changed. That is, he first tells us that the chief observes the bravery of the warriors, then does not dance, and this causes the braves to have been cowardly. This raises the memory problem. In the example presented, what actually would have been the case was that the chief observed the cowardly behavior of his braves but mistakenly believed he had observed their being brave. (Gorovitz talks as if the chief observes their being brave.) What must be explained is not a change in memory, but rather why the chief makes the mistake in observation. This is not such a bizzare situation as might at first seem, since the mistake in observation might explain why the chief did not dance for the last two days. Further, as was discussed when considering Gale's argument, given that the chief believes he knows whether the warriors were brave, then it follows that he will not deliberate about either bringing about or not bringing about the warriors' bravery. Consequently, when Gorovitz raises the problem of the chief not being able to rely on his own observation, he is not quite on the mark. In those cases where the chief observes what goes on, he will not later attempt to change anything.

After considering the case in which the warriors were brave, Gorovitz turns his attention to the possibility that the chief observes the warriors' cowardice. Here, considerations parallel to those for the brave warriors are raised—what about the chief's memory if he later dances and causes the young men to have been brave, etc? Here too, similar objections can be raised as was done for the objections to the chief witnessing bravery: the warriors were not first cowardly and then brave, etc.

Because Gorovitz considers the problem of explaining the agent's change in memory a crucial one, he considers one way out. A restriction might be placed on backward causation cases such that they occur only when the agent is ignorant of the prior event in question. (It has been argued that this will, in fact, be the case, following from the notions of *deliberation* and *intention.*) Thus all one must worry about is some outside observer. Gorovitz says, however, that here there is a similar problem. "If *anyone,* let alone the agent, witnesses the nonoccurrence of E at T, then we are ... unable to account for the error or change in *his* memory in the event that the agent by performing A at T causes E to have happened" (p. 369). Here again, one can point out that there is no change in what was the case at a given time, T_1, in the past.

From what has been seen of Gorovitz's arguments, it is not surprising that he concludes that talking about backward causation leads to absurdity. This is so, he claims, because of the way *causality* is related to *memory* and other related concepts. From what was said above, it should also not be surprising that the conclusion is rejected.

Pears on Dummett

In "The Priority of Causes,"[10] D. F. Pears comments upon the argument presented by Black and Flew against Dummett. The general position Pears takes is that these arguments (as was suggested earlier) are not telling, and if one is to reject Dummett's thesis, then one will have to move to different ground.

10. David F. Pears, "The Priority of Causes," *Analysis* 17(1956-57): 54-63.

First Pears discusses the objection to Flew and Black's criticisms. Since this was done earlier, it will perhaps be best to go directly to what Pears offers as new objections to Dummett's position that certain conditions could incline one to adopt the concept of *quasi-cause*. It might be argued that it is contradictory to suppose that a concept like effectiveness can be applied to subsequent conditions. When discussing such considerations, Dummett might argue that even if this were so, it would only show something about the existing conceptual system. What is really at issue is, Why could not the relations between these concepts be rearranged? Pears has two reasons such a rearrangement is impossible.

The first is that, if "effective" means sufficient and/or necessary and prior,

> . . . then it may be no use looking for some further characteristic in virtue of which a sufficient and/or necessary condition is effective if and only if it is prior. . . . There must be a limit to the unearthing of mediating characteristics, and the limit is set by the number of the series of related concepts. In this case the number seems to be two. [P. 59]

There are several ways one might argue against Dummett's first objection. The most direct would be to deny the first premise and point out that there is more involved in the notion of *effective* or *cause* than *sufficiency* and *necessity*. This will be discussed in a later section on what is involved in the notion of cause, and for this reason the detailed discussion of Pears' first objection to backward causation will be postponed.

The second point is that priority is an indivisible concept. That is, there is nothing which can be detached from the concept and still have what one might wish to call "priority." There are two ways one might argue against this rather sketchy objection to backward causation. First, one can point out that Pears has offered no evidence to support his claim that priority is an indivisible concept. Second, it seems as if he is aiming at the wrong concept. When discussing the relationship between "effectiveness" and "priority" in the hope of presenting a palatable case for backward causation, one is not looking for an analogue of priority which has one aspect of priority severed while all other features remain. Rather, one wishes to keep the notion of *priority* intact while disassociating it from that of *effectiveness*. Thus, even if Pears' claim of the indivisibility of priority were correct, it would not be to the point.

Pears also considers a rather indirect approach to Dummett's thesis. Dummett presented the case of the man who uses L to bring about E, and who regards L as effective only when he does not know about E. To justify the effectiveness depending upon ignorance, Dummett says that the effectiveness of L is not a fact of nature like L itself. When the man knows about E, it is reasonable to view L as ineffective; and if he does not know E and views L as effective, then it is.

Pears is quick to point out that the contrast between what is a fact of nature and what it is reasonable to believe is not very clear. If all *reasonable* means is *reasonable on the available evidence*, then the man would not be using a new conceptual system, rather merely misusing one. When he added to his knowledge to learn about E, he would thus correct his mistake, and it would then be unreasonable to view L as effective. Thus Dummett cannot have this distinction in mind, though he does not suggest a viable alternative.

Although Pears's criticism seems well directed, it is not telling to Dummett's backward causation thesis. This is so because Dummett has accepted a fallacious argument. It does not matter if the man knows E or does not. A later action can be effective. One can know E, and E can have been the case because of his subsequent action. This is logically possible. What would be the case (not what could be the case) is that once one knew E, he might say there is now no need to do L. The consequences of what happens after he makes this decision (he accidentally does L, etc.) has been discussed earlier. An analogue for forward causation may help make this point. A person may precognize that he will shoot himself and then, at some later time, pull a trigger and indeed shoot himself. Here one would not want to say that the pulling of the trigger was ineffective. The precognitive knowledge does not make the shooting ineffective, nor should it cause one to believe it is ineffective—similarly with backward causation.

At the end of his essay, Pears presents a good formulation of a general difficulty with backward causation, specifically with any attempt to present a scheme in which both backward and forward effectiveness operate. Because they are incompatible (*E* and *L* cannot be linked by both backward effectiveness and forward effectiveness), there must be some way of preventing them from linking the same pair of events. Another way of putting the difficulty is to ask, What properties differentiate events linked by forward effectiveness from those linked by backward effectiveness? It should immediately be pointed out that if the question is legitimate, it does not show that the difficulty is insurmountable, formidable as it may be. It might also be argued that the question is improper, and there does not in fact have to be any distinguishing properties between the differently linked pairs of events, just as there are no distinguishing properties between balls which are tossed up in the air and those hurled to the ground. Some happen to have been thrown upward and others downward.

III

Swinburne and Affecting the Past

Knowledge is the object of our inquiry, and men do not think they know a thing till they have grasped the 'why' of it (which is to grasp its primary cause).

Aristotle (*Physics*, Book II, Ch. 3)

Swinburne is probably unique among the various writers on the backward causation question. He has focussed his attention on one small aspect of the problem which no other writer considers as deeply and applies his comments on this aspect to Dummett's dancing Indian chief. Thus it will perhaps be appropriate at this point to see what Swinburne has to say on the matter.

In his first paper on the topic,[1] he makes a very important distinction, and it is this distinction which is his contribution to the backward causation question. In a trivial sense one can affect the past. What a person does now can determine what is a correct description of the past. The example used is, "The bomb dropped on Hiroshima was the last one dropped by man." In this example, what happens after the bomb is dropped will determine what a correct description of the situation is. What one must do to tackle the problem of

---

1. R.G. Swinburne, "Affecting the Past," *Philosophical Quarterly* 16(1965): 341-47.

affecting the past is to distinguish these trivial cases from those which are of philosophical interest. It is here that Swinburne has his helpful suggestion. The interesting cases concern those statements about the past which, in principle, can be more conclusively verified at t_1 than at some later time, t_2. This criterion rules out statements such as the Hiroshima one because there is reference to the future. Such a statement can better be verified at a date later than that on which the bomb was dropped.

Swinburne considers the example offered by Dummett and then makes a general point. If, after the fourth day of the chief's dancing, the observers report that the men were not brave and the chief then dances and does affect the past (and makes the men to have been brave), then the observers will change their report. The chief will have changed the evidence also, and the observers will appear to have lied at first, or to have been mistaken, etc. In any event, the evidence at the later time will be that which is accepted. According to Swinburne's criterion, this case is trivial. In general, if one is to affect the past, then one will also affect the evidence, and consequently it seems as if evidence at a later time is that to be accepted, not of the earlier time. Thus the event here has a future component (according to this criterion) and is not one of the kind which is of philosophical interest.

Here it will be necessary to examine Swinburne's criterion in detail to see precisely what it is, and where Dummett's example fails to meet it.

> Statements apparently about a past time which an observer could have found out at the time to be true or false, had he been correctly positioned and executed tests, and which he could have verified or falsified more conclusively at the past time referred to, had he been correctly positioned and executed all possible tests, than he could have done at any subsequent (or prior) time, are indeed statements solely about that past time. Statements apparently about a past time which an observer, had he been correctly positioned and executed all possible tests, would not have been able to verify or falsify at the time referred to more conclusively than at a subsequent time are statements with a covert future reference. [P. 341]

This criterion is granted. What now remains to be seen is where the Indian chief example goes off the course outlined by the criterion. After presenting the example, Swinburne makes a general comment:

> Any defender of the thesis of the possibility of affecting the past must claim, as Dummett does, that the action alleged to affect the past produces changes in the evidence about what happened in the past. . . . For unless this happens, there would be no grounds for supposing that the action had any effect in the past at all. . . . [P. 344]

This seems false. For a counter example one can even consider Dummett's example where the chief does not know the results and dances on the last two days. One can imagine a correlation between dancing and bravery as strong as is wished. Here, there does not seem to be the need for a change in evidence. When the observers return they merely report what happened. Although there is no change in evidence, there is reason for believing that the chief affected the past.

Even when discussing cases where the bravery or non-bravery is already known, it does not seem as if a correct description of what occurs must involve a "change in evidence." Swinburne says, "The change in evidence may therefore consist either, as in Dummett's example, of a change in the memory claims, or of a change in the traces" (p. 345). Here, two possibilities are discussed. First the observers may have made a mistake and been poor observers. If after the chief dances their error is discovered, this does not mean that their memories have been erased and replaced with new ones. The memories of the braves' cowardice *must* be compatible with the new evidence—there is no logical contradiction involved; they made a mistake, and mistakes are explainable. Thus rather than say the evidence "changed," one might more accurately say that new evidence was uncovered, and this evidence suggested a different conclusion.

There is a parallel situation in the case of the observers

lying, rather than making a mistake. Here too, after the chief's dance the evidence is not changed. Rather, new evidence turns up showing that the warriors were brave and that the observers were lying.

What is not clear from the above discussion is just where Dummett's example fails to meet Swinburne's criterion. As the example is set up, the evidence which *in fact* is best, and which is accepted, is that of the later time. But this does not violate the criterion. *In principle* the best evidence still could have been obtained at the time of the hunt, had all possible tests been made. (Swinburne offers nothing to suggest why this is not the case.) Perhaps the reason Swinburne feels the criterion has been violated is because of his assertion that all cases of affecting the past must involve changing evidence. But this is false.

In his summarizing remarks he says, "Now the observer may believe that they lied, that they knew all along that the young men had been brave. In which case they would have no grounds for saying that the chief's dance had affected the past; the only effect of the chief's dance was to make them now tell the truth" (p. 346). Just why there are no grounds for saying that the chief's dance affected the past is not clear. Perhaps there is a confusion of *affecting* and *changing* the past, but this would be especially surprising, since Swinburne notes this distinction earlier in his paper.

From what was said earlier, it is clear that one must reject Swinburne's conclusion that the statement "the young men were brave" is not solely about the past but at least partly about the future. Consequently one must also reject his more general conclusion that no change in evidence could ever substantiate the claim that someone had affected the past.

There is a rather different approach which might help eliminate some of Swinburne's confusion. One might attempt to formulate a simpler criterion for distinguishing philosophically trivial from nontrivial cases of affecting the past.

Consider this criterion: A past event at t_1, if affected by the present, will be of philosophical interest if and only if that event can be observed at that time (t_1). Consider the following event: The priest baptized the world's greatest pianist. At the time of the baptizing, t_1, the observers could indeed see a baby being baptized and may have had reason to report that this is what they saw. In any event, they could not have had reason to report that they observed a pianist being baptized. Thus on this criterion, this particular case (a priest baptizing a pianist) has reference to the future, and affecting it (the eighteen-year-old boy deciding to become a pianist) will not be of philosophical interest. This criterion does suggest that there are interesting cases of affecting the past from which no logical contradictions are immediately derivable.

In his book on space and time, Swinburne devotes a chapter to the problem of affecting the past.[2] In this section entitled "Past and Future," he presents what are basically the same arguments as were seen in his earlier article. Since there are some different formulations of the position however, it might be instructive to examine some of these statements with a view toward seeing precisely why Swinburne comes to the conclusion he does.

When discussing why Dummett's example is not of philosophical interest, Swinburne offers a formulation which reveals where he goes astray.

> But the point is that so long as an agent (or any observer) could (logically) have found out at t_1 what happened with greater certainty than he could at t_3, then the truth or falsity of any statements describing what happened is independent of the agent's subsequent action and any consequences thereof. Only if the agent or any observer of his action could not (logically) have found out what happened with greater certainty, is it plausible to suppose that the truth or falsity of such statements could be affected by the action. But then such statements will not be really about the past. [Pp. 166-67]

---

2. R.G. Swinburne, "Past and Future," in *Space and Time* (London: Macmillan, 1968), pp. 157-71.

Here Swinburne seems to be suggesting that as long as a statement has a truth value at time t_1, then nothing subsequent to t_1 can have any effect on that truth value. It is true that if at t_1 a statement has a value of T, then nothing can be done to subsequently cause that statement now to be F. But this is merely the old question of *changing* the past. Swinburne is also ruling out, however, the possibility that the reason the statement at t_1 has a truth value of T is that at a later date something was done which caused the event described by the statement to have occurred. But to merely rule out this possibility, as Swinburne does, is to rule out precisely what is at issue—the possibility of affecting the past.

It might also be pointed out that an analogous argument might be given with regard to future events which would entail determinism. That is, if there is an event in the future, a statement of which will have a certain truth value, then nothing can be done now to have an effect on that truth value; consequently nothing that is done now can have any effect on the event in question.

The last section of the chapter "Past and Future" is based on the assumptions that Swinburne has demonstrated the logical impossibility of affecting the past. Using this assertion, he discusses why certain models of the universe are not possible. Since the premise is not granted, there is no point in examining the argument.

There is still another, more direct avenue by which Swinburne may be attacked. It will be remembered that the discussion began with a consideration of a proposition, the truth of which depended on a later event. Then the question was raised as to whether such an instance is a trivial or nontrivial case of backward causation. Here the objection might be raised that the question of backward causation does not even arise. Merely that the truth value of a proposition is determined by a later event does not entail that backward causation is an issue. What would have to be demonstrated is that whenever the truth value of a proposition is determined by a later event, there is a causal connection between the later event and the event or state of affairs described by the proposition. This question is even broader than the one of backward causation. Even when the truth value of a proposition describing a state of affairs is determined by an earlier event, one still might not want to say that there is a causal connection between the state of affairs and the prior truth-determining event. Thus for Swinburne to firmly establish his point, he would have to supply a missing demonstration.

IV

Scriven and Flew

*Now although it be true, and I know it well, that
there is an intercourse between causes and effects,
so as both these knowledges, speculative and opera-
tive, have a great connexion between themselves
yet because all true and fruitful natural philosophy
both a double scale or ladder, ascendent and de-
scendent, ascending from experiments to the invention of causes, and descending from causes to the
invention of new experiments.*

Bacon (Advancement of Learning,
Second Book, VII. 1.)

In all of the papers discussed so far, those writers who
positively assert the backward causation thesis merely argue
for the logical possibility of an instance in which a cause
comes after its effect. In "Randomness and the Causal Order,"[1] Michael Scriven goes one step further and attempts to
show that there are actually cases in which one is practically
forced to accept that backward causation has occurred. The
example Scriven relies upon to demonstrate this is an ESP
experiment conducted by Whateley Carrington over thirty
years ago.

The subjects in this experiment are asked to draw what

---

1. Michael Scriven, "Randomness and the Causal Order," *Analysis*
17(1956-57): 5-9.

they believe the experimenter will hang on his wall the following day. After the subjects have made their drawings, a randomizing device (a roulette wheel, opening a table of random numbers, etc.) determines to what page a dictionary will be opened. A drawing is then made of a noun on that page; the randomizing device also determines which noun is chosen. After the drawings are received, the target picture is then hung, and judges apply an objective set of criteria to see if the drawings correspond to the target. It can be assumed that there is a very high statistical correlation between the drawings and the target picture. This is a situation in which Scriven feels one should talk in backward causation terms.

He says that so far as the perception model is concerned, one would want to say that Tuesday's picture influenced Monday's drawings. This is so because with ordinary perception one says that the object perceived causes the response to the perception of it. Thus one might say the target hung causes the drawings of it. Here, of course, the question may be, and should be, raised as to whether or not the perception model is appropriate. Several arguments can be presented against it. First it might be stated that perception entails a fairly constant faculty. In the case of visual perception, for example, when a person has his sight tested, unless something radical happens to his eyes, he expects his sight to remain near the same level for a while. He certainly does not normally expect his vision to be adequate on rare occasions and totally absent on most. Such inconsistency, however, seems to be the case with extrasensory perception. Further, when ESP tests are given, the level of scoring seems to be related to the number of alternatives from which the subject is attempting to guess the target. If one is choosing from five alternatives, his rate of success may be 0.25 instead of the 0.20 expected by chance. If he then changes and attempts to guess which of ten alternatives is the target, his rate of success may change to something like 0.15. It seems as if the number of alternatives affects the number of correct perceptions one has. With visual perception, this is not the case. When one can clearly see what a drawing is, it does not matter from how many others it has been selected. Thus if one establishes a level of vision (say he can distinguish half of all the letters he is shown), it will not matter if additional alternatives are added to the pool from which targets are selected. His level of correct identification will be expected to stay the same.

The above two arguments against the appropriateness of the perception model for ESP are not suggested as conclusive. They merely show that objections can certainly be raised, and Scriven is going to need firmer ground for establishing backward causation.

Scriven's next defense of his position rests upon the random selector. It might be said that Monday's drawings determine the Tuesday drawing, but the random selector blocks this criticism. This is so because it is the random selector which determines which drawing is to be hung. The obvious argument is that perhaps the Monday drawings affect the randomizer. Scriven feels, however, that this would involve staggering difficulties and enumerates them.

(a) If many people make similar drawings on Monday, how can it be explained if they do not have the common cause of the Tuesday drawing? This is such a weak objection that it seems hardly worth a reply. Why can there not be a prior common (perhaps psychological) cause of the commonness of the drawings? Scriven overlooks this when he says, "To account for clustering, we would presumably need some brand-new form of ESP at the unconscious level" (p. 6).

(b) The next objection is, How can such a cluster affect the random selector since the spatial and physical relations are absurdly unlike any in known cases of influence? Here it seems as if Scriven is asserting that backward causation is the more plausible of the two alternatives with which he is dealing. Again, there is no defense of this position, and it is merely presented as self-evident.

(c) Scriven's third argument is again an appeal to the

improbability of affecting the random selector. The question he raises is, How can the random selector be affected in such a way that the code used to determine the item to be drawn will produce the drawing corresponding to the cluster? Granted it would be a radically new kind of cause, but then surely so is backward causation. Which is the more radical is not so clear as Scriven seems to feel it is.

(d) Granted that perhaps some sort of PK can influence random selectors such as roulette wheels, dice, etc., what about the case where a random number table is the selector? There are many arguments against this position. Perhaps the best is merely to point out that if the selection principle is an experimenter arbitrarily opening a table of random numbers and then these random numbers determining the drawing, then there is an alternative. The experimenter could use his clairvoyance of both the previous drawings and the random numbers to select a favorable entry point into the random number table. Indeed, there is even a small amount of experimental evidence for such a phenomenon.

All of Scriven's arguments against the drawings affecting the target selector involve an appeal to implausibility. To show how superficial such an approach is it is only necessary to quote his invective against PK affecting the selector. "This is pretty exciting stuff, even if a trifle hazy as hypotheses go; but not very obviously parsimonious by comparison with a little reversal of the causal order when we add the hypotheses required to deal with (a), (b), and (c)" (p. 6).

Since they have somewhat similar ways of doing philosophy, it is perhaps appropriate that Flew has published a direct comment[2] on Scriven's article. As was seen, the randomizer is the focal point of Scriven's argument. Flew comments upon this, but in a way different than here.

In Scriven's example, the "causes" are the drawings which

---

2. Antony Flew, "Causal Disorder Again," *Analysis* 17(1956-57): 81-86.

the experimenter hangs, and these are determined by the randomizer. Flew makes the point that all other causes are levers, or in principle can be used as levers, to bring about their effects. By having the randomizer determine the drawings, Scriven has placed the causes beyond human manipulation, and thus they cannot be used to bring about their effects. Here Flew is sound in pointing out the reverse restrictions which Scriven places upon his proposed example of backward causation. (From the earlier discussions, it is evident that there is no need for such limitations.) Flew goes on, however, to make stronger claims. "Scriven's cunning stipulation must sterilize necessarily all the usual implications of efficacy, his suggested description of his example in terms of causation reversed is seen as at worst contradictory and at best pointless" (pp. 82-83). Packed into this sentence are two assertions which require close attention. The first is the claim that if an event (the drawing hanging) is beyond human manipulation, then it cannot be said to be efficatious. The second implied assertion is that the possibility of such human manipulations is necessary for an event to be called a cause. The first assertion is false. When one knows for certain that an event was not brought about by human manipulation, he does not refrain from saying that it was efficatious. For example, when a star burns itself out and life on a nearby planet dies, one says that the dying on the planet was a result of the star's demise. Concerning the second point, there is nothing peculiar in speaking of non-human-manipulated causes, so why should there be difficulty in talking about causes in which the possibility is removed? If the example were one of forward causation, one would not be hesitant to call the hung drawing a cause. That is, let us imagine that the randomizer selects a drawing, and after it is hung, art students look at it and draw their own pictures, which are later shown to have a significant statistical correlation with the one hung by the experimenter. Certainly there is nothing wrong with saying the drawing selected by the randomizer is the partial cause of the art students' drawings which cluster on the same topic. Thus this human manipulation (or the possibility of it) does not seem essential to the ascribing of causality to an event.

The point Flew raises does, however, call attention to the necessity for a somewhat detailed account of what is involved in the notions of *cause* and *effect*. It is with this approach that the later sections will deal.

V

Mundle, Broad and Ducasse: Is the Problem with the Notion of Precognition?

> There are things too not a few for which it is not
> sufficient to assign one cause; you must give sev-
> eral, one of which at the same time is the real
> cause.
>
> Lucretius (Nature of Things, Bk. IV, 703)

Mundle

In an article entitled "Does the Concept of Precognition
Make Sense?"[1] C. W. K. Mundle comes to the conclusion
that the concept does not make sense, and alternative explanations must be found. One virtue of Mundle's treatment is
that he attempts to analyze what is entailed by the term
"precognition" as it is normally used.

Before the philosophical discussion, there is a presenta-
tion of the evidence of precognition. It is not necessary to
enter into such a discussion here, since it may be assumed
that the evidence is as strong as can be imagined. Thus begins
the philosophical portion of the paper.

Mundle says that "precognition" and "ESP" are loaded
with theory.

---

1. C. W. K. Mundle, "Does the Concept of Precognition Make
Sense?" *International Journal of Parapsychology* 6(1964): 179-98.

They have been chosen on the assumption that the facts in question are to be assimilated to sense-perception, and that just as sense-perception involves knowledge by acquaintance, so the psychic subject must possess knowledge by acquaintance, knowledge of objects or events not accessible to his senses. [P. 181]

He goes on to assert that if "precognition" is defined in terms of non-inferential knowledge of a future event "one will have to say that the knowledge in question is usually unconscious and that it spasmodically influences the subject's conscious experience on actions" (p. 182).

Mundle seems to believe that there is something wrong with this, but never states precisely what it is. It would appear that what Mundle rejects is a fair description of the situation. In any event, Mundle prefers causal terminology. Thus, the psychic subject can be described as responding appropriately to events which he has not observed or inferred. Eventually, Mundle rejects this description also because it would involve "a cause-factor influencing what happened earlier." This, he says, is self-contradictory because part of the meaning of "cause" is that a cause must precede its effects. He supports this by saying that causal explanations involve mechanisms whereby causes modify structures of substances, initiate chains of events which span the spatio-temporal gaps between causes and effects, etc. Mundle then asserts that it seems inconceivable that what occurs now could be acted upon by things which are not yet in existence. This is why Mundle rejects the possibility that precognition involves a cause coming after its effect.

To summarize Mundle's position, it may be said that he has argued that the notion of precognition does not make sense for two reasons: (1) It involves the concept of knowledge by acquaintance which would be unconscious and spasmodic, and there seems to be something wrong with this. (2) It involves the concept of a cause succeeding its effect, and there is something self-contradictory about this.

One sees nothing objectionable in the first "objection."

Concerning the second, it was argued earlier that there is no reason to believe that it is logically impossible for a cause to come after its effects. Thus Mundle's second reason for rejecting precognition can have merely linguistic import, if it is indeed part of the meaning of "cause" that the cause precedes its effect. Thus, one might not reject precognition, but merely come up with a revised definition of "cause" or a new word to use in place of the old.

Since he has rejected precognition, Mundle is still faced with the task of explaining the facts in nonprecognitive terms. One alternative he offers is unconscious experience. That is, all ostensible precognitive experiences are the result of acquiring knowledge by nonprecognitive means and then inferring the future event in question. Mundle's second alternative involves the "precognitive experience" causing the precognized event to come true. It is not necessary to discuss either of these attempts to discredit the concept of precognition, for Mundle is correct in saying that they do not involve precognition. This paper is concerned with the logical possibility of precognition, not alternative explanations of the facts or evidence for precognition.

Broad

A treatment of the problem somewhat similar to Mundle's is offered by Broad in a recent article, "The Notion of Precognition."[2] Broad, who has probably done more work on the concept of precognition than any other philosopher, gives what is perhaps the most thorough analysis of "precognition," or "ostensible precognition," as he calls it. Rather than defining the word "precognition," Broad chooses to define the statement "X was a precognition of Y." This definition contains five clauses. A summary of the essence of each follows:

---

2. C.D. Broad, "The Notion of Precognition," *International Journal of Parapsychology* 10(1968): 165-96.

(1) X was either a single or sequence of human actions, and Y a single event or sequence of events.

(2) X happened at t, and Y at a later moment, t_2.

(3) Y corresponds in detail to X in such a way that one should say that Y was a fulfillment of X.

(4) It is not a mere coincidence that X was followed by Y.

(5) Y's being related to X cannot be accounted for by either (a) X being a cause-factor or causal ancestor of Y or (b) X and Y having a common cause-factor ancestor, W.

There are three points which should be made concerning Broad's analysis, two of which can be made relatively quickly. The first point is merely that for some reason Broad seems to believe that precognition is an ability only of humans. This is clear from his first clause of his definition. There is no reason for this restriction, and indeed there are many cases reported of apparent precognition in animals. Should Broad's remaining four conditions be satisfied, there is no reason that he should not accept the case as an example of precognition. This is a minor point, however.

The second point is similar to one discussed in relation to Mundle's article and involves Broad's fifth condition. Here, Broad also rules out the possibility that both the precognition and the precognized event have a common cause prior to the precognition. Once again, it must be pointed out that this is merely a linguistic point and not a logical one. Broad, like Mundle, has merely legislated what he wishes to consider as a precognitive experience. In terms of the evidence available, there seems no reason for ruling out such a possibility. Consequently, although this consideration might eventually be shown to be one of a set which suggests that it is impossible to obtain evidence for precognition as Broad defines it, it in no ways suggests that the overt evidence which parapsychologists claim to have collected involves contradictions, and thus there must be some error.

The third point involves the issue which has been central in this section: Can a cause come after its effect? Broad clearly denies the possibility and asserts:

> ... the alternative suggestion, viz., that the *fulfilling event or state of affairs* contributes to set up a chain of effects and causes which contributes to cause the *pro-referential experience*, is plainly nonsensical. For *until* the event which will answer to the present experience in such a way as to be a fulfillment of it, shall have happened, *nothing* can be caused by it. And *when* it shall have happened, anything that it may contribute to cause must be later than it. [Pp. 191-92]

Broad does not offer a demonstration for his claim; he merely presents it as a self-evident truth. He does point out that, in the case of a cause coming before its effect (as in a case of past-perception, or seeing an event which is in the past, such as viewing sun-spots), the past event is connected with the present perception by a causal chain. In this chain each· event is an effect of the immediate successor in the chain. According to Broad, nothing analogous is possible in the case of precognition. This is so because, until an event will happen, nothing can be caused by it. The point here is not merely a linguistic one, arising from the way "cause" and "effect" are used. Rather, it is a factual point. A future event is nothing but an unrealized possibility and thus cannot influence anything. Broad goes on to claim that it is probably this fact which has caused "cause" and "effect" to be used in the way they are.

Although Broad is begging the question by asserting what is at issue, his treatment still has instructive value. This is so because he explicates how the impossibility of a cause succeeding its effect mitigates the notion of *precognition*. There is no condition in the definition which asserts that the precognized event must be the cause of the precognition of it. The problem arises when the fifth condition is examined for X being a precognition of Y. Here, he has ruled out X causing Y, and X and Y having a common cause prior to each. To this Broad adds the self-evident fact that Y cannot have caused X. Thus it seems as if X and Y can in no way be causally related, and coincidence is left as the explanation for the correspondence between X and Y. This, however, is ruled out by the fourth condition in the definition which says that the relation between X and Y must not be mere chance coincidence. From what has just been stated, Broad's conclusion is not surprising: As he defines it, nothing can satisfy his definition of precognition. From what was stated earlier when considering the questions of backward causality, opposition to Broad's position is evident. His first error lies in admitting the possibility of Y causing $X;$ this is something which he has prematurely rejected. Second, there is also the possibility of the common prior cause.

Throughout the analysis of causality, one works within the framework of the constant conjunction conception of causation. That is, A and B are said to be causally related if they are constantly conjoined. This thesis, of which Hume is the foremost proponent, has many problem areas. For example, night and day always follow each other, but is one the cause of the other? It will be noted that this statement of the thesis allows for backward causation. This is not actually the case as Hume puts it. One of his rules by which one is to judge causes and effects is that the cause must be prior to the effect.

Ducasse

There have been many criticisms of Hume's position, but few philosophers have attempted to offer usable alternatives. In *Causation and the Types of Necessity*,[3] C. J. Ducasse attempts this. For purposes here, most of Ducasse's thesis can be ignored. He asserts that causality is a relation obtained among three terms of a perfect experiment: A, the state of affairs, and B and C, two changes in it. It will be of interest

---

3. Curt John Ducasse, *Causation and the Types of Necessity* (New York: Dover Publications, 1969).

here to briefly note what Ducasse has to say about backward causation.

In discussing Russell's theory of causation, Ducasse points out that one result of what Russell says is that there is no reason to assume that a cause cannot come after its effect. Ducasse feels this is wrong and says:

> But I say that the only conclusion which is open to any English speaking, or rather English understanding reader, when he is told that an effect may well be supposed to precede its cause, is either that the assertion is staringly false, or else that in spite of appearances, it is not really expressed in English, but in some other tongue in which the words "cause" and "effect" also occur, but surely with very different meanings indeed than in English. [P. 42]

Although Ducasse's causal theory does not specifically legislate against backward causation, Ducasse feels compelled to make the linguistic legislation against it which has been seen several times before.

It is interesting to note that Ducasse has an article which deals specifically with how his notion of causality relates to parapsychology.[4] The article is general and does not, for the most part, deal with precognition. Rather, Ducasse briefly recapitulates his triadic theory of causation and discusses determinism, psychokinesis, etc. Toward the end of the article he mentions the causal question with precognition and sketches the form his solution will take: He will try to make out that "past," "present," and "future" can be defined only with reference to psychological states and have no meaning in purely physical terms. Ducasse suggests that such a realization will help one out of the apparent paradox involved in precognition experiences suggesting a cause later than its effect. He does not give any of the details of his argument here, but presents them more fully in a later article.[5]

----

4. Curt John Ducasse, "Causality and Parapsychology," *Journal of Parapsychology* 23(1959): 90-96.

5. Curt John Ducasse, "Broad on the Relevance of Psychical Research to Philosophy," in *The Philosophy of C.D. Broad* (New York: Tudor Publishing Company, 1959), pp. 375-410.

In this paper, Ducasse is criticizing a paper in which Broad tries to show how parapsychology is relevant to philosophy.[6] In this paper Broad discusses a wide range of topics, and Ducasse comments on each of these. This section of the discussion is limited to the precognition-causation paradox. Broad's position in the article is basically the same as in his later article in the *International Journal of Parapsychology* which was discussed earlier. He argues that it is an analytic truth that no event should have effects before it has happened. What is of interest here is that Ducasse accepts this "analytic truth" and attempts to show that it is not violated by precognition. However, some other, more vulnerable assumption must yield in the resolution of the paradox.

The basis of Ducasse's theory is that no definition of "past," "present," and "future," applied categorically, can be given in purely physical terms. Thus physical events, apart from the psychological events which are percepts of them, are not categorically past, present, or future. Physical events are past, present, or future only conditionally—relative to other physical events. Ducasse gives a rather confusing explication of what he means here, but it may be possible to avoid analyzing this portion of his thesis. However, what he says about psychological events must be considered.

Ducasse claims that "past," "present," and "future," which apply only to psychological events, derive their meaning from a characteristic—*liveness*—which is present to different degrees in the psychological events. As an example, the hearing of the word "inductively" is presented. When the whole word is heard, liveness is that characteristic which is possessed in its maximal degree by the syllable "ly," in somewhat lower degree by "tive," etc. This is the definition offered for "liveness."

For several reasons this "definition" is not sufficient. First,

---

6. C.D. Broad, "The Relevance of Psychical Research to Philosophy," *Philosophy* 24(1949): 291-309.

it seems as if there might be several characteristics which "ly" has in the maximal degree, "tive" to a lesser degree, etc. Here, other characteristics may be included, such as being later in time. If other characteristics are mentioned, Ducasse might reply by saying that liveness includes *all* such characteristics, but with being later in time, he cannot do this, since concepts like "present" are inseparably tied to time. But there is more wrong with the definition than just this.

Consider the psychological event of watching and being engrossed in a movie. While the viewer is attentive to the dialogue, someone whispers "inductively." It would seem as if the experience of hearing the word "inductively" has less liveness than the experience of hearing the movie dialogue, both of which are contemporaneous. This is a situation Ducasse would not want.

Still another difficulty with the definition is that "present" is defined in terms of *maximal* liveness, but no criterion for determining maximal liveness is offered. At first it seems as if it might be difficult to determine what is present. Another way of putting this is that the difference between past and present, as Ducasse makes it, is one not of quality, but of quantity. The only difference between a past and present event is that the present one has more of a certain characteristic. Because of this, there is a special difficulty in defining "future." If "present" means having the most liveness and "past," having less liveness, what is to say about "future"? In defining "future" Ducasse says nothing positive about it; he presents a negative definition. All psychological events which are neither past nor present are future. This too leaves something to be desired. Before offering his example involving "inductively," Ducasse said that he could not describe the characteristic about which he was talking. This may suggest that there might be something basically wrong, or at least incomplete, about his treatment of "past," "present," and "future" as predicates of psychological events.

In summarizing his thesis, Ducasse says, ". . . psychological events can be *actually experienced* only in the order of their respective degrees of liveness, and thus of recency. . . . This entails that *the time-series of psychological events has intrinsic direction*" (p. 389). The first point to be made here is that the connection between liveness and recency has not been established as Ducasse says it has. It seems quite possible for an earlier event to have more liveness than a later event. One remembers his last birthday more vividly (inferring this is related to liveness) than the day after it. Thus, liveness is not necessarily concurrent with recency. From this emerges a second point concerning the "intrinsic direction" of psychological events. It is conceivable that a person's experiences could exhibit an inverse relation between recency and liveness. Ducasse has given no argument against such a possibility. Thus there seems to be no entailment that psychological events have intrinsic time direction.

It is probably evident how Ducasse uses his theory to resolve the causal paradox involved in precognition. Physical events, by themselves, do not categorically have pastness, presentness, or futureness. When a given physical event is precognized (and precognition of it is a psychological event) then the physical event receives presentness, not before. Thus there can be no problem of a nonexistent event causing something to be in the present. As soon as an event is precognized, it is present.

Aside from attacking Ducasse's method of definition, there are several approaches one can take. The following is one of the less obvious objections. An essential feature of Ducasse's scheme is that physical events, by themselves, do not have futurity. Thus he can solve the problem of precognizing physical events. He cannot, however, handle the precognition of purely psychological events—events which would have pastness, presentness, etc., independent of the precognition of them. An example is precognitive telepathy, the precognition of someone's thoughts.

VI

Taylor, Chisholm and Dray: The Philosophy of Language

Now sir, as to the point that Tullius clepeth 'causes,' which that is the laste point, thou shalt understande that the wrong that than hast receyved hath certeine causes, whiche that cleches clepen Oriens and Efficiens, and Causa longinqua and Causa propinqua; this is to seyn, the fer cause and the ny cause.

Chaucer (Tale of Melibeus, par. 37)

When discussing the question of whether or not a cause can come after its effect one might wish to settle the matter by purely linguistic considerations. That is, it is possible that by analyzing the meanings of "cause" and "effect" one might discover that included in the meaning of these two words is a prohibition of a cause succeeding its effect. If this is found to be the case, then the question as to whether a cause can follow its effect is conclusively settled in the negative. If, however, it is discovered that there is nothing in the meaning of the two words which prohibits such a relationship, then the question *may* be answered in the affirmative, but not necessarily so. That is, there may be other reasons that a cause cannot come after its effect, apart from the meanings of the words.

In principle one should be able to decide rather easily whether or not the proposition, "A cause cannot succeed its effect," is an analytic truth. This, of course, is done by consulting a dictionary. In practice it is quite a difficult task, and philosophers do not agree on this simplest of issues.

In "Why Cannot an Effect Precede Its Cause?"[1] which we discussed briefly earlier, Max Black attempts to show why it is necessary that an effect succeed its cause. Professor Black's main approach is that of the philosopher of language; he wishes to show that what is meant by "cause" and "effect" precludes the possibility of saying that an effect can precede its cause. Black considers the question posed in the title of his paper as of essentially the same kind as, Why cannot Tuesday come before Monday?—it is necessary. Such questions, he claims, reduce to questions of semantics and why certain words cannot be used in certain ways. It should be noted that Black in no way attempts to analyze the meanings, uses, etc., of "cause" and "effect." Rather, he merely analyzes "Monday" and "Tuesday" and then asserts that "cause" and "effect" constitute a similar case. With preliminaries out of the way, Black then continues his theme by saying, "Since we are to examine a necessary statement ('An effect cannot precede its cause'). . ." (p. 52). This is an assertion for which Black has laid little groundwork.

Mundle is another philosopher who asserts that it is an analytic truth that a cause cannot succeed its effect. Mundle[2] rejects several descriptions of precognition because they would involve "a cause-factor influencing what happened earlier." This, he says, is self-contradictory because part of the meaning of "cause" is that a cause must precede its effects. He supports this by saying that causal explanations involve mechanisms whereby causes modify structures of substances, initiate chains of events which span the spatio-temporal gaps between causes and effects, etc. Mundle then asserts that it seems inconceivable that what occurs now

---

1. Black, *op. cit.*
2. Mundle, *op. cit.*

could be acted upon by things which are not yet in existence. This is why Mundle rejects the possibility that precognition involves a cause coming after its effect.

Black and Mundle are not the only philosophers who argue for the analyticity of the statement, "A cause cannot succeed its effect," but they will suffice as examples of proponents of this position.

In his article in *The Encyclopaedia of Philosophy*[3] Flew is concerned with the cause-effect problem raised by precognition; namely, that the effect seems to precede its cause. Thus, if a person precognitively calls cards in correct sequence, the cause of his response may be said to be the cards, even though they do not yet exist in the sequence which he precognized. Flew claims that it is a "necessary truth" that a cause must either precede or be simultaneous with its effect. He does not offer much in the way of establishing the necessity of his truth. Because he chooses to start from the assumption of this "necessary truth," Flew is forced into a rather curious position regarding precognition: that it is all concidence. The road he takes which leads to this dead end should be considered.

The possible causal relations between two events A (a series of calls) and B (a series of cards) are as follows: (1) A results from B; (2) B results from A; (3) A and B result from a third cause; and (4) A and B result from independent causes. The first possibility is ruled out by Flew's "necessary truth." The second (which could be PK) and third possibilities are ruled out by the definition of *precognition*. What is left is the fourth proposition, which is precisely what is meant by *coincidence*. In the discussion of Black's similar conclusion earlier, the weakness of such a position was pointed out.

---

3. Antony Flew, "Philosophical Implications of Precognition," in *The Encyclopaedia of Philosophy*, Vol. 6 (New York: Macmillan, 1967), pp. 436-41.

Flew tries to strengthen his thesis by pointing out that to establish a statistical correlation is not to establish a causal connection. Although this is true, one wonders if many readers will be satisfied with the conclusion Flew has left for them.

One might ask here, if Black, Mundle and Flew were correct, what would the effect of their arguments be? Since theirs is a linguistic one, the import must be linguistic. That is, under certain conditions (e.g., when a certain member of a relation comes temporally after the other member) one cannot use the words "cause" and "effect" to describe the members. Further, it might be pointed out that even if it were an analytic truth that a cause cannot come after its effect, new evidence sufficiently striking might arise which would make one want to give up the analytic truth. Thus the more telling analysis in the backward causation question might be whether or not it is possible to conceive of circumstances under which one would relinquish the analytic truth. However, even to do this one must clarify the meanings of "cause" and "effect." This is precisely what Taylor and Chisholm attempt to do.

In their paper, "Making Things to Have Happened,"[4] Chisholm and Taylor do not attempt to solve the question of backward causation; they merely wish to show that the basic arguments against backward causation are incorrect. To do this they try to clarify the notions of "sufficient" and "necessary" causes. Their definitions are as follows:

(1) "A is sufficient for B" means: (A and $\sim B$) is impossible.

(2) "A is necessary for B" means: ($\sim A$ and B) is impossible.

"Impossibility" is undefined here.

Taylor and Chisholm then use the notion of *sufficiency* to define "causation." "A causes B" means A is sufficient for B. They point out that neither A nor B designate a single event,

---

4. Chisholm and Taylor, *op. cit.*

rather a set of events. Thus a match being struck is not the cause of the flame because it is not impossible that it be struck and no flame appear. (It could be wet.) However, there is a set of conditions which is sufficient for the flame: the match is dry; there is oxygen; it is struck; etc. This set of conditions, being sufficient for the flame, *causes* the effect in question.

The authors then point out that, "Now surely there is nothing in the causal relation, other than this relation of sufficiency" (p. 74). To get the match lighted, all that is required is a certain set of sufficient conditions. One does not need something like "power," "effectiveness," etc., to guarantee the effect. Since all that is involved in causation is sufficiency, then it is clear that a set of sufficient conditions coming after an event could certainly be considered as the cause of the prior event.

Chisholm and Taylor's conclusion is basically correct—that a set of sufficient conditions coming after an event can be the cause of the event—but the means by which they arrive at the conclusion is questionable. Specifically, they are surely wrong when they assert that there is nothing in the causal relation other than sufficiency. At the risk of sounding like an iterated modality, their position can be put as follows: A sufficient condition for a set of circumstances being a cause of an event is that the circumstances comprise a sufficient condition for the event in question. What this writer asserts is that it is a *necessary*, but not sufficient, condition for being a cause that a set of circumstances comprise a sufficient condition for a particular event. Translated into Taylor and Chisholm's phrasing, this means there is something in addition to sufficiency that makes up the causal relationship. An example or two will illustrate this.

Consider the case in which there is a reporter who always reports accurately what he sees. Given this, then whatever the reporter reports can be said to have been caused by him. This is so because the situation has been set up in such a way that it is impossible for the event reported not to have occurred and there to be a report of the event. This is the criterion (or definition of) sufficiency and, in turn, of causation. It is possible that the reporting causes the event to have occurred, but it does not seem as if one must call the reporting the cause, as Chisholm and Taylor seem forced to do. Another way of pointing out the deficiency in their definition of "cause" is to realize that, if sufficiency is the causal relation, given several sets of conditions, each sufficient for event *E*, there is no way of deciding which is the cause.

Dray takes a somewhat similar stand[5] when he points out that the Taylor-Chisholm criterion of cause does not allow one to make the distinction between events which are merely *signs* of other events and events which are *causes* of other events. Given that there is a properly-operating barometer (which is a sign of the rain to come), one would want to be able to refrain from calling it a cause of the rain. In such cases there is a common cause prior to both the sign and the event heralded by that sign.

Dray is correct in pointing out that the definition of "sufficient" as presented is not strong enough to cover causation. When it is said that a cause is *sufficient* for its effect, this is stronger than saying that *A* is *sufficient* for *B*. There is something involved in the causal relationship beyond the ordinary logical sense of *sufficient*.

---

5. William Dray, "Taylor and Chisholm on Making Things to Have Happened," *Analysis* 20(1959-60): 79-82.

VII

Mackie:
Toward a Revised Notion of Causation

Felix, qui potuit rerum cognosceri causos. (Happy the man who has been able to know the causes of things.) Virgil (Georgics, II, 490-91)

At least one omission in the discussion of backward causation should be obvious. What constitutes a cause and what constitutes an effect have not yet been adequately defined. In pointing up errors made when attacking this problem, Taylor, Chisholm and Dray in the preceding chapter showed that the notions of *sufficiency* and *necessity* were not adequate. In a recent paper,[1] J. L. Mackie attempts to specify what delineates a cause from an effect.

Mackie begins by pointing out that, although temporal priority may often be used to determine which of two events is the cause and which is the effect, there seems to be more of a difference than mere temporal priority. Indeed, the fact that a cause and effect are sometimes spoken of as being simultaneous, coupled with the fact that it seems logically possible that a cause could come after its effect, suggests that one may be able to describe a relation such as *causal priority*

---

1. J.L. Mackie, "The Direction of Causation," *The Philosophical Review* 25(1966): 441-66.

between two events, and the relation will not be identical with nor entail temporal priority.

After discussing some criteria which are obviously not satisfactory, Mackie presents the case for *effectiveness* as the distinguishing characteristic between causes and effects. In a two-membered relationship, the causally prior item is the one which can be directly controlled, and by means of which the other is indirectly controlled. Thus in the case of striking a match and producing a flame, the cause is the striking of the match, since one uses the rubbing of a match against a rough surface to bring about the flame, but does not use a flame to bring about the rubbing of a match.

Mackie suggests that a basic problem with this criterion is anthropomorphism. The question may arise as to what decides causal priority in a world without voluntary agents. This is so because causes have been delineated from effects in terms of *use* and *control,* and these are primarily anthropomorphic terms. In a world without inhabitants, it would seem as if there would be no use or control, but one might still wish to say that there were causal relationships. Also, it would seem that one should be able to describe causal priority merely in terms of a relationship existing between the two events in question. The notion of *effectiveness* as defined does not allow this. Thus "effectiveness" will not be sufficient for the purpose at hand.

At first it might be objected that the statement, "*A* lends itself to being controlled through *B,* but not *vice versa,*" describes the causal priority of *A* over *B* solely in terms of the relationship between *A* and *B*. What this criticism suggests is that some explication is needed for delineating properties which are explicable solely in terms of two members of a relation and properties which require external considerations. As a preliminary criterion the following is suggested: A property will be said to be explicable solely in terms of two members of a relation if and only if, were there nothing existing other than the members in question, the property could still be seen by an external observer to be predicable. Thus in the case of the preliminary criticism, the dispositional term "lends itself to" requires knowledge of externals, since at least possible volitional agents who control things must be considered. As a relational property which can be seen to hold between two members without reference to anything other than the two members, the relation *larger than* can be considered. If only two spheres existed, the external observer could still see that *A* was larger than *B*, even if there were no rulers, volitional agents, etc.

In an attempt to illuminate the problem, Mackie discusses a standard case of what has been construed as an experiment providing evidence for backward causation, the drawing experiment. This is the experiment involving an artist on Monday trying to produce a copy of a drawing which will be hung on Tuesday. What drawing will be displayed on Tuesday is still undecided and will be decided by a random process. It may be assumed that precautions against cheating and other normal explanations for a correspondence between the precognizer's drawing and the one hung on the following day are adequate. It may also be assumed that the correspondence between the two drawings is perfect, or nearly perfect, and that this is the case repeatedly. Can this experiment yield evidence to support the hypothesis that Tuesday's drawing is causally prior to the drawing completed on Monday?

The argument for backward causation rests to some extent upon the exclusion of all known means of Monday's drawing influencing the one hung on Tuesday. It is the exclusion of these possibilities that might lead one to accept backward causation as the explanation. This, of course, is a radical explanation, but if forced, then one must accept it. However, there is another radical explanation—that there is a mysterious unknown way in which the temporally prior drawing influences its successor. Thus if the alternative is excluded, something must be accomplished other than the exclusion of known ways the earlier drawing can affect the later one. There are two ways of doing this.

First, one might resort to an examination of the random procedure to show that the determination of the second drawing is indeed random and thus beyond the influence of the earlier drawing.

Second, one could trace the causes of the second drawing, *B,* to some event, *C,* which is prior to and causally independent of the earlier drawing, *A.* With this second alternative, there is an immediate problem, the question of whether *C* is indeed causally independent of *A.* Here, the evidence must be weighed and a decision made as to which is more plausible, backward causation or a mysterious forward causation. Thus this suggestion will not enable one to settle the question decisively.

In favor of accepting backward causation as the path of least resistance, it might be suggested that a person is familiar with perception in general and knows how it is caused. Consequently when asked for the mechanism explaining the correlation between the drawings, he has some reply. In the case of mysterious forward causation there is no answer. For this reason, backward causation is the better of the two alternatives. However, it can be objected that all experience with sense perception is in terms of forward causation. To argue that the drawing example is a familiar case is certainly stretching a point. Further, it has not been established that extrasensory perception is analogous to sense perception in the relevant way which would be essential for the above argument to be sound. Thus if the drawing experiment example is to suggest an acceptable case of forward causation, it will have to be argued on grounds other than an analogy between extrasensory perception and sensory perception.

It might be argued that backward causation is less puzzling than mysterious forward causation, but this is a rather tenuous position, since all backward causation is mysterious as far as a mechanism for explaining the correlation goes. Consequently, it could be asserted that, since all familiar cases of causation are cases of forward causation, less conceptual revision is involved in postulating a mysterious forward causation.

Concerning the first alternative, one might suggest that there is always the possibility that the earlier drawing affects the random procedure, so that one cannot rule out causation in the ordinary temporal sequence. For this objection one can specify that the random procedure is such that it can be statistically checked to see that it is indeed random. Thus if there were a causal connection between the earlier drawing and the process for selecting the second drawing, the process would not be truly random, and this could be detected. Thus one can specify that he is discussing only those cases in which the process is truly random.

Here it might be argued that, since the procedure is random and there is a one-to-one correlation between items selected by the procedure (target drawings) and the first drawings, then the first drawings must also be random in some sense. Since there is a correlation between two sets of drawings, each of which is said to be random, then it is just as easy to say that the first set caused the second as to say that the second caused the first. Thus the specification of the random procedure merely gives the appearance of helping to settle the issue in favor of backward causation—actually it contributes little.

Mackie has a different objection to the drawing example. It is the one Black offered earlier. In this case, A is already fixed, and B is not yet decided. Mackie says,

> ... But on every occasion, after the drawing is made, it is possible that someone or something should intervene so that the corresponding pattern fails to be produced. Consequently, it cannot on any occasion be the pattern that is responsible for the details of the drawing; the precognition hypothesis must be false even for those occasions when the device is not stopped and the pattern actually is produced and turns out to be just like the drawing. [P. 455]

Here the point is that since the cause, *B*, can always be avoided (and the effect, *A*, remains), then one cannot say that *B* is in fact the cause of *A*.

The argument, however, is vacuous. When *B* is avoided, precognition is no longer the issue; when *B* cannot be avoided, *A* is a sufficient condition for *B*. However, the crucial cases when *B* is not avoided but could have been do not entail that *B* cannot have causal priority over *A*. It is apparent that had *B* been avoided, then *A* would not have been the case; but allowing *B* to happen causes *A* to have been the case. The inference is not a valid one in the above argument, but it is accepted by Mackie and forces him to look for an escape.

His escape leaves him with only those cases in which, at the time of *A*, there already exists a sufficient cause of *B*, which is causally independent of *A*. This, however, rules out the first alternative—that the precognized drawing is un-caused in the sense of being the product of a random pro-cess—and thus Mackie must say such events cannot be precog-nized in the sense he means. The only possible case of backward causation for him is that involving the cause, *C*, of *B* which is temporally prior to *A* and causally independent of it. This is a case in which there is no conclusive evidence of backward causation, but evidence must be weighed in order to choose which explanation of the correspondence between drawings is preferred.

To summarize Mackie's position thus far, he has concluded that there can be no decisive experiment for the backward causation hypothesis. This is tentatively agreed. However by accepting an invalid argument, he has unnecessarily limited the cases in which backward causation is possible. Specifical-ly, backward causation does seem a logical possibility when *A* is fixed and *B* is still unfixed. (An event at time t is fixed if and only if either the event has occurred at or before t or a sufficient cause of the event has occurred at or before t.)

This analysis of backward causation is central to defining

"causal priority." It leads Mackie to suggest the following: ". . . if *A* and *B* are causally connected in a direct line, then *B* is causally prior to *A* if there is a time at which *B* is fixed, while *A* is not fixed otherwise than by its causal connection with *B*," (p. 457). From this it should be clear that Mackie has ruled out any case of straightforward backward causation as it has been discussed; this is so because the past is fixed, and earlier Mackie has stated that because the "effect" has either already happened or not happened, it cannot be brought about. As argued earlier, this is not necessarily true. It seems logically possible for one to cause something in the past to have happened by a present action.

Mackie believes, however, that he has left room for a kind of backward causation which is indirect. For example:

> . . . While a backward causal relation could not be used on its own to bring something about, it is conceivable that it should be used in conjunction with a forward causal relationship. . . . Someone could use *C* to bring about *A* by means of *B*. By doing something on Sunday to ensure that a particular pattern was produced on Tuesday, one could bring it about that a corresponding drawing was made by the precognizer on Monday. So although what was brought about (the drawing) necessarily (in view of the fixity of the past) occurs later than the bringing of it about, *part* of the chain of causation by which it is brought about might be time-reversed. [P. 458]

For two reasons, the above example is not clear. The first reason for the confusion is revealed by the first sentence quoted. Mackie is asking, How can a backward causal relation be used? What is at issue is whether a backward causal relation is possible. He seems to feel that of course it is possible—the only question is how it is used.

The first difficulty leads to a second which concerns just what Mackie's example entails. It might rightly be asked whether Mackie has presented anything suggestive of backward causation. All he has said is that: (1) Something on Sunday causes something on Tuesday; and (2) This causes something on Monday to occur. Clearly this is merely a stock case of forward causation. The question now is, Precisely what is it that causes the Monday drawing? It seems as if Mackie is saying that it is the conjunction of the Sunday and Tuesday occurrences. That is, Sunday's doings and Tuesday's events are sufficient to bring about the Monday drawing. Thus since Tuesday's event is a necessary condition for an earlier event, this is a case relevant to the backward causation question. Here it might be pointed out that, because Sunday's activities are sufficient to cause Tuesday's patterns and because Sunday's activities combined with Tuesday's patterns are sufficient to bring about Monday's drawing, then it can be said that Sunday by itself is a sufficient condition for Monday's drawing. Given Sunday's activities, Monday's drawings follow. Thus the example presented is not a clear-cut case of backward causation.

With this point made, Mackie goes on to discuss priority with respect to the dispersal of order. It is in this section that the essence of causal priority comes closest to explanation. The argument of the dispersal of order is a general one attempting to show that the arrow of time must go in the direction traditionally depicted; and as a consequence of this general truth, it can be seen that causality also goes in one direction. The argument makes use of a specific kind of phenomena such as entropic processes. There is the well known example of a stone being thrown into a placid pool of water, and circular waves spreading out from the center being observed. Although the reverse process—concentric waves contracting toward a center—is a logical possibility, it is never observed. If it were, it would be difficult to explain. One is used to explaining a number of coherent separated items (the waves) by referring to a single event (the dropping of the stone) causing them. If concentric waves were seen converging at a center, it would seem as if this would call for a third event which causes these waves to converge as an explanation. Thus there seems to be a direction of explanation, from a central event to coherent separated events. Thus it might be said that explanation runs to events involving dispersal of order from events which do not. The above is a sketch of a general argument showing why there is a direction of explanation with respect to the dispersal of order and suggesting why one takes the arrow of time as going in the traditional direction. It must now be shown how this relates to the direction of causation.

It is generally accepted that a cause explains its effect in a way in which an effect does not explain its cause. This is so even when the effect can be used to retrodict the cause. Given this, it is not unreasonable to hope that by considering how the explanatory powers of causes and effects differ, insight might be gained into the nature of causal priority. It might first be noted that when a question is asked by saying, "why?" the answer generally begins with "because." That is, when one is asking the why of things he is asking for the cause; it is the cause which explains why its effect is. Clearly the effect does not explain why its cause is. All effect can do, if it is a sufficient condition for its cause, is guarantee that the cause occurred. This is not an explanation. (There will be a later discussion of how this relates to this writer's explanation of causal priority.)

Consider the case of a heavy metal ball resting on a soft cushion. Here is a case in which a cause and its effect are simultaneous, and temporal priority will not help to decide which is the cause and which is the effect. It would almost certainly be argued that the downward force of the ball causes the depression in the cushion, as opposed to arguing that, when the ball is placed on the cushion, a depression appears and causes the ball to fall. The reason the first alternative is chosen is that taking the downward force exerted by the ball as the cause explains the depression. If the depression were taken as the cause, there would be need of an explanation as to how the depression came about. This example suggests that the kind of explanatory power a cause has is explicable in terms of bringing about its effect. This relationship is asymmetrical. This will be discussed in more detail toward the end of this section, when a preliminary criterion of causal priority is presented.

Attention can now be directed to Mackie's final conclusion, which in spite of the errors noted above, is completely agreeable. Causal priority is something other than temporal priority, and the fact that the two may always coincide is merely contingent. Should there be an instance in which it seems as if they do not coincide, then backward causation can be a coherent explanation. However, the evidence for backward causation can never be conclusive and must be weighed against the inductive evidence for all causation being forward.

Although a point of agreement with Mackie can be found, his solution to the causal priority question is rejected. From the discussion of the explanatory power that a cause has which its effect does not have, it would seem as if the solution to the crucial question is what Mackie rejected earlier—effectiveness. That is, in a causal relationship, the member which can be controlled to indirectly bring about the other is the cause. The objection which Mackie raised concerning anthropomorphisms is not insurmountable. Even in a world without volitional agents, the notions of *effectiveness* or *use* can be utilized to distinguish a cause from an effect. It could be said that the member which one would use to bring about the other is the effect. Here Mackie would point out that causal priority is being made to be something not merely in terms of a relationship between two events in question, and this suggests something is wrong.

Earlier, in attempting to crudely explicate what is involved in such basic relational properties, the example of two spheres, one larger than the other, was used. Here "larger than" could be seen as appropriate, even if the two spheres were the only entities in their universe. It should be clear that most properties would not satisfy the criterion suggested. For example the notion of *sister of* cannot meet this criterion. To make sense of *A* being the sister of *B,* one must be able to refer to parents. There are similar requirements for many other relational concepts. What Mackie does not explain, and what might be questioned here, is why causal priority must be a concept which can be discussed solely in terms of the relationship existing between the two members in question. Indeed, it might be argued that when dealing with members so complex as events, such atomic explications of relations are impossible.

Something which Mackie discusses very early in his article would even imply this. When discussing the case of a house fire started by a short circuit, Mackie makes the point that although one would say that the short circuit was the cause of the fire, by itself it is not even a sufficient condition for the fire. There are other conditions which are necessary, such as the wood near the short circuit being dry, etc. This might lead one to conclude that the cause-effect relationship is not dyadic, but perhaps triadic. In *Causation and the Types of Necessity,*[2] Ducasse attempts to establish the point that there is a triadic relationship between the cause, effect, and various boundary conditions. There is no need to discuss here whether Ducasse's view is correct. This is merely intended to show that Mackie's assumption that the cause-effect relationship is some sort of atomic dyad is far from established.

Another case in which there are simultaneous causes and effects (PV = kT) may help show why external considerations are necessary. In the case of gas in a rigid container, one can keep the volume of the gas constant and manipulate the temperature so as to increase the pressure. In this case, the cause is the temperature manipulations and the effect the increase in pressure. The pressure could also be increased and thus increase the temperature. Here the cause and effect roles are reversed. Thus given two variables, temperature and pres-

---

2. Curt John Ducasse, *Causation and the Types of Necessity* (New York: Dover Publications, 1969).

sure, one cannot determine which is the cause and which is the effect without external considerations.

That is, if all that existed were the gas in the container, one would not be able to distinguish between the cause and the effect. However, once it is realized that there is nothing illegitimate about bringing in external considerations, then it becomes possible to distinguish a cause from its effect. In the case of the pressure and temperature, the one which is manipulated, or might be manipulated, to bring about the other is the cause. From this it is seen why a cause has an explanatory power which an effect lacks. The cause is that which is used to bring about the effect; thus it holds the answer to the "why?" of the effect.

It might be argued that *control,* or *effectiveness,* cannot be the key to causal priority, since in the example of pressure and temperature, either could be manipulated to bring about the other. Thus, how can "control" distinguish the cause from the effect? All this shows is that either could be the cause and either could be the effect. In any specific situation, only one is the cause and the other the effect; but the one which was manipulated to bring about the other is the cause.

It is perhaps relevant to note here that the Greek word for cause, $\alpha\iota\tau\iota\alpha$, is one taken from the vocabulary of the law courts. There it was used to denote where the blame lay. That is, the word used to point out the responsibility for (bringing about) the injurious action later became the word used in common speech to denote a cause in general. This too perhaps suggests the essential bond between a cause and effectiveness.

VIII
Schlesinger and a Causal Criterion

---

*Present events are connected with preceding ones
by a tie based upon the evident principle that a
thing cannot occur without a cause which produces
it.* Laplace (Essay on Probabilities, Ch. II)

In·a recent unpublished paper,[1] George Schlesinger suggests a criterion for distinguishing a cause from its effect
which is in line with the suggestion in the last section. It will
perhaps be instructive to consider this paper in some detail.

Schlesinger begins his paper by making some general remarks about the backward causation problem. Contained in
this first section is an example which is quite helpful in
demonstrating an important point. Frequently it is asserted
that, if two events are causally related and the temporally
prior one is necessary and sufficient for the later event, then
the prior event must be the cause. Schlesinger offers God's
omniscient knowledge as a counter-example. Here is the case
that God's knowledge of a later event is both necessary and
sufficient for the event to occur, but His knowledge is not
construed as the cause of the events. Rather, the situation
might be viewed as the events causing God to have His
knowledge of them. This example shows that an event may
be both necessary and sufficient and prior to a second event

---

1. George Schlesinger, Untitled, unpublished manuscript (Chapel
Hill, N.C.: University of North Carolina, 1970).

and still not be the cause of the subsequent event. This clearly shows that, if there is a criterion for distinguishing a cause from its effect, it will have to be other than temporal priority.

The second section of Schlesinger's paper presents three possible examples of backward causation which Schlesinger feels shed light on the nature of causal relationships in general. Of the three, the first comes closest to giving an example of a case more favorable to a backward-causation explanation rather than a forward-causation explanation.

The example involves a magician in the remote Himalayas, cut off from direct contact with the outside world. He hears reports of outside events days or months after they occur. The sorcerer is particularly interested in hurricanes and tornadoes. Whenever he hears that one has occurred, he names an amount of money and performs unusual rites to ensure that the damage caused by the storm is equal to the amount of money he named. Repeatedly it is the case that the amount of damage equals to the nearest hundred dollars the amount of money named by the magician. Here, the question can be asked whether this suggests that backward causation is the best answer. That is, does the magician by the performance of his rites cause the damage to have been what it was? The alternative, in terms of forward causation, is that the prior event, the amount of damage occurring, causes the magician to name the given amount. As the example is given, there is no advantage in selecting the explanation involving backward causation; both explanations are mysterious in that neither offers a mechanism connecting the causes and effects. However, since all experiences with cause involve the cause preceding the effect, this alone is the deciding factor and suggests that the forward-causation explanation is the one to be accepted.

Suppose that to the example described is added the fact that scientists have been told of the magician and come to investigate him. They bring radios with them and listen for reports of storms. When they hear the amount of damage, they ask the magician to name a different amount and perform his rituals. In every case the scientists later are told that the first report was in error (because of errors in addition, inaccurate reporting, etc.), and the correct total is precisely that named by the magician. Now there are two phenomena to explain. Again there is the correlation between the naming of an amount and the actual amount of damage. But what explains the curious misreporting? It is here that a backward-causation explanation might be chosen because then there is an explanation for the misreporting. The scientists' behavior is the cause of the misreporting, and this behavior is explained by their desire to thoroughly test the magician. If the misreporting caused the scientists to ask the magician to say the particular sums, then there still is no explanation for the misreportings. Thus, Schlesinger concludes that here is a case in which one might wish to choose backward causation as an explanation rather than forward causation.

It should be pointed out that this example illustrates the close connection between causation and explanation. If forward causation is selected as the explanation, there is still need of an explanation for the occurrence of the misreportings. However, because the backward causation has more explanatory power it is selected as the more desirable.

Schlesinger's main contribution occurs in the third section of his paper, where he presents a case in which A and B are mutually necessary and sufficient. Is it possible that A occurs, but an indeterministic event prevents B from happening? (It might at first be asserted that this is impossible by the definition of "sufficient." However, what is meant by "sufficient" is that *within the laws of nature* if A is sufficient for B and B obtains, then B must occur. In the example, an indeterministic event is outside the laws of nature.) Schlesinger's answer is most illuminating and suggests a criterion for distinguishing a cause from its effect:

... it depends, whether A is the cause of B or its effect! If A is the cause, then its occurrence ensures by nature's laws that B is going to occur. In spite of this, it may yet happen that B fails to materialize, namely if something indeterministic that is something outside nature's laws prevents B from occurring has no influence on A's happening, which is not causally dependent on B and, as long as whatever the cause of A has taken place, A will occur. But if B is the cause of A, then it is immaterial whether the failure of B to occur was in accordance with the laws of nature or came about by a chance event; A cannot occur, since its necessary cause has not taken place. [Pp. 8-9]

The last part of the last sentence quoted above is objectionable. That is, A can occur; all that is needed is a cause outside the laws of nature. Indeed, this is almost certainly what Schlesinger intended. Here a necessary cause is only necessary within the laws of nature. The important point is that if B is the cause of A, *either* a natural or unnatural event blocking the occurrence of B also blocks the occurrence of A. If B is the effect, then so long as A's cause has taken place, A will occur.

Although the test for delineating a cause from its effect is a good one, one might ask if it is necessary or sufficient. That is, can all cases be put to this test, or will it suffice for only some? Here the answer is that it is sufficient but not necessary. One can imagine circumstances in which the test cannot be used to decide which of two causally related events is a cause and which is an effect. In the test, one is asked to imagine one of the two events being prevented from occurring. This is necessary for the criterion to get off the ground. This should suggest the limitation for the test. Usually it is decided which of two causally connected events is the cause and which is the effect without imagining the consequences of an indeterministic event preventing one from occurring. Surely a criterion is desired which would cover cases in which one does not have to imagine one of the two events not occurring. It is true that Schlesinger's test will suffice in many instances, but because of the limitation just mentioned, it cannot be construed as shedding light on the nature of cause in general. It simply does not cover the circumstances under which it is normally decided which event is the cause and which is the effect.

It should be pointed out that Schlesinger's suggestion is merely a test to be used to determine which of two events is the cause. Thus even if universally applicable, it still might not elucidate the meaning of "cause." For this reason, the suggestion made at the end of the preceding chapter is perhaps superior and logically prior to Schlesinger's. What is meant by "cause" is elucidated by the notions of *effectiveness* and *use*. What is meant by a cause is that which can be used or is effective in bringing about something, and this suggests, as a test, asking which of the two members of a causal relation would be (or is) used to bring about the other. This is a universally applicable test and perhaps does more to get at the meaning of "cause."

It may be noted that Schlesinger's criterion does rely upon one aspect of the notion of *effectiveness*. That is, the way in which the criterion distinguishes between a cause and its effect is by delineating different kinds of effectiveness. If effectiveness is lost by the introduction of either a natural blocking or indeterministic blocking of A's occurrence, then A is the cause of B. If effectiveness is lost only through indeterministic block of A, then A is the effect of B.

Conclusion

That there is nothing in the effect, that has not existed in a similar or in some higher form in the cause, *is a first principle than which none clearer can be entertained.*
Descartes (Reply to Second Objections)

The first section of this monograph dealt with a para-psychological example and the last ended with a proposed method for distinguishing a cause from its effect. Such a strange journey deserves some recapitulation.

The main purpose of this paper was to discuss and at least tentatively settle the question of whether a cause can come after its effect, and consequently whether this is a possible way of describing precognitive experiences. There was an examination of the arguments and examples of the philosophers who believed they had demonstrated the logical impossibility of a cause succeeding its effect, and in each case the arguments were found in some way to be deficient. Thus it can be said that so far there has not been adequate proof that a cause cannot come after its effect. Frequently errors arose because a distinction was not made between changing the past (a logical impossibility) and affecting the past. This, however, was not the main source of confusion; that has a more basic cause.

Perhaps the main problem was that the concept of *cause* had not been clearly defined, and consequently there was no test available to delineate a cause from an effect. That such an important philosophical concept has not been clearly explicated may strike the reader as surprising, but it should not. It might be asked, What other concepts have been so explicated that there is now no room for philosophical arguments concerning them? Clearly, the answer is, None. Most concepts are so intricately connected to others that an attempt at complete explication must inevitably lead the philosopher of language on a chase throughout the entire language. In the discussion of cause it was necessary to tie that concept with those of explanation, control, effectiveness, etc. To render a complete explication of "cause" would have required explication of the other terms just mentioned, and of course this is impossible in a paper of this length, or for that matter, of any finite length. This does not suggest that one should not attempt to continually make clearer the ways in which various concepts are used. It does imply that the suggestions concerning the concept of *cause* must be taken as tentative.

An attempt has been made to explicate the meaning of "cause" by reference to the concepts of *effectiveness* and *use,* and this has been done with the help of the notion of *explanation.* This immediately suggested the use of these concepts as criteria for delineating which of two causally related members of a pair is the cause and which is the effect. Thus the one which might be controlled to bring about the other, or the one which explains the other's existence in a nonreciprocal way, is the cause. Since the terms used and the criteria suggested are atemporal, it is clear that the door to the possibility of a cause succeeding its effect is left open.

Thus the investigation of backward causation leads to the even bigger question as to the meaning of "cause" in general. It is hoped that the attempted solution to this backward causation problem is at least a minor contribution and that the preliminary clarification of the notion of *cause* will form the basis for a further, and perhaps more important, contribution.

Only one point remains to note, and that is the relevance of the discussion of backward causation to the logical possibility or impossibility of precognition. As has been seen, it was frequently asserted that precognition is a logical impossibility because it entails backward causation, which is logically impossible. Thus the settling of the backward causation question has relevance to the field of parapsychology. Since examination of the arguments against backward causation found none to be adequate, even if precognition did necessarily involve backward causation, it cannot be ruled out on the grounds that backward causation is a logical impossibility. It might also be pointed out that in the incidental treatment of this topic an adequate demonstration that precognition necessarily involves backward causation was not seen.

Bibliography

Ayer, A. J. "Why Cannot Cause Succeed Effect?" In *The Problem of Knowledge,* pp. 170-75. Baltimore, Md.: Penguin Books, 1957.

Black, Max. "Why Cannot an Effect Precede Its Cause?" *Analysis* 16(1955-56): 49-58.

Chisholm, Roderick M., and Taylor, Richard. "Making Things to Have Happened." *Analysis* 20(1959-60): 73-78.

Dray, William. "Taylor and Chisholm on Making Things to Have Happened." *Analysis* 20(1959-60): 79-82.

Broad, C. D. "The Notion of Precognition." *International Journal of Parapsychology* 10(1968): 165-96.

Broad, C. D. "The Philosophical Implications of Foreknowledge." *Aristotelian Society Proceedings* supp. 16(1937): 177-209.

Broad, C. D. "The Relevance of Psychical Research to Philosophy." *Philosophy* 24(1949): 291-309.

Broad, C. D. "A Reply to My Critics." In *The Philosophy of C. D. Broad,* pp. 774-86. New York: Tudor Pub. Co., 1959.

Ducasse, Curt John. "Broad on the Relevance of Psychical Research to Philosophy." In *The Philosophy of C. D. Broad,* pp. 375-410. New York: Tudor Pub. Co., 1959.

Ducasse, Curt John. "Causality and Parapsychology." *Journal of Parapsychology* 23(1956): 90-96.

Ducasse, Curt John. *Causation and the Types of Necessity.* New York: Dover Pub., 1969.

Dummett, M. A. E. "Bringing about the Past." *Philosophical Review* 23(1964): 338-59.

Dummett, M. A. E. "Can an Effect Precede Its Cause?" *Aristotelian Society Proceedings* supp. 27(1954): 27-44.

Feather, Sara R. and Brier, Robert. "The Possible Effect of the Checker in Precognition Tests." *Journal of Parapsychology* 32(1968): 167-75.

Flew, Anthony. "Can an Effect Precede Its Cause?" *Proceedings of the Aristotelian Society* supp. 27(1954): 45-62.

Flew, Antony. "Causal Disorder Again." *Analysis* 17(1956-57):81-86.

Flew, Antony. "Effects Before Their Causes?—Addenda and Corrigenda." *Analysis* 16(1955-56): 104-10.

Flew, Antony. "Philosophical Implications of Precognition." In *Encyclopaedia of Philosophy*, vol. 6, pp. 436-41. New York: Macmillan Co., 1967.

Gale, Richard M. "The Impossibility of Bringing About the Past." In *The Language of Time*, pp. 103-33. New York: Humanities Press, 1968.

Gale, Richard M., ed. *The Philosophy of Time*. Garden City: Doubleday & Co., 1967.

Gale, Richard M. "Why a Cause Cannot be Later than its Effect." *Review Metaphysics* 19(1965): 209-34.

Gorowitz, Samuel. "Leaving the Past Alone." *Philosophical Review* 23(1964): 360-71.

McCreery, Charles. *Science, Philosophy & ESP*. Hamden, Conn.: Archon Books, 1968.

Mackie, J. L. "The Direction of Causation." *Philosophical Review* 25(1966): 441-66.

Mundle, C. W. K. "Does the Concept of Precognition Make Sense?" *International Journal of Parapsychology* 6(1964): 179-98.

Pears, David F. "The Priority of Causes." *Analysis* 17(1956-57): 54-63.

Rand Corporation, The. *A Million Random Digits with 100,000 Normal Deviates*. New York: The Free Press, 1966.

Rao, K. Ramakrishna. *Experimental Parapsychology*. Springfield, Ill.: Charles C Thomas, 1966.

Reichenbach, Hans. *The Direction of Time*. Edited by Maria Reichenbach. Berkeley and Los Angeles: University of California Press, 1956.

Rhine, J. B. *Extra-Sensory Perception*. Boston: Boston Society for Psychical Research, 1934.

Rhine, J. B. and Brier, Robert, eds. *Parapsychology Today*. New York: Citadel Press, 1968.

Rhine, J. B. and Pratt, J. G. *Parapsychology*. Springfield, Ill.: Charles C Thomas, 1957.

Rhine, Louisa E. "Precognition and Intervention." *Journal of Parapsychology* 19(1955): 1-34.

Schlesinger, George. Untitled, unpublished manuscript. Chapel Hill, N.C.: University of North Carolina, 1970.

Scriven, Michael. "Randomness and Causal Order." *Analysis* 17(1956-57): 5-9.

Smart, J. J. C., ed. *Problems of Space and Time*. New York: Macmillan Co., 1964.

Swinburne, Richard G. "Affecting the Past." *Philosophical Quarterly* 16(1965): 341-47.

Swinburne, Richard G. "Past and Future." In *Space and Time*, pp. 157-71. London: Macmillan Co., 1968.

Whiteman, Michael. *Philosophy of Time and Space*. New York: Humanities Press, 1967.

Index

Subject Index

Name Index